PERFORMATIVE ARCHITECTURE

BEYOND INSTRUMENTALITY

EDITED BY BRANKO KOLAREVIC & ALI M. MALKAWI

Spon Press
Taylor & Francis Group
NEW YORK AND LONDON

First published 2005
by Spon Press
29 West 35th Street, New York, NY 10001

Simultaneously published in the UK
by Spon Press
11 New Fetter Lane, London EC4P 4EE

Spon Press is an imprint of the Taylor & Francis Group

Printed and bound in Spain by Grafos SA

British Library Cataloguing in Publication Data
A catalogue record for this book is available from the British Library

Library of Congress Cataloging-in-Publication Data

Kolarevic, Branko, 1963-
Performative architecture : beyond instrumentality / Branko Kolarevic and Ali Malkawi.
p. cm.
Includes bibliographical references and index.
ISBN 0-415-70083-3 (hardcover : alk. paper)
ISBN 0-203-53685-1 (ebook)
1. Buildings--Performance. 2. Buildings--Energy conservation.
3. Architecture--Decision making. I. Malkawi, Ali. II. Title.
TH453.K65 2004
720--dc22

2004014235

CONTENTS

ACKNOWLEDGMENTS

We are indebted to the contributors of this book for the time and effort they have devoted to writing the chapters and for putting up with the various editorial demands.

We are grateful to the sponsors of the symposium "Performative Architecture," which was held at the University of Pennsylvania (Penn) in October 2003, and which led to this book. The symposium was organized by the Digital Design Research Lab (DDRL) and the Building Simulation Group (BSG) in the School of Design at Penn (PennDesign). It was sponsored by McGraw-Hill Construction, Turner Construction, Autodesk, Bentley Systems, Flomerics, the Department of Facilities and Real Estate Services, the Department of Architecture at Penn and by PennDesign.

We would like to thank the University of Pennsylvania and the School of Design for providing an exceptional environment for teaching and research. We are particularly grateful to Professor Gary Hack, Dean of PennDesign, Professor Detlef Mertins, Chair of the Department of Architecture, and Professor Richard Wesley, former Head of the Department of Architecture, for their support of our endeavors. Finally, we want to thank our students and our colleagues for the stimulating conversations about the subject of this book.

PROLOGUE

BRANKO KOLAREVIC

PROLOGUE

BRANKO KOLAREVIC

This book discusses an emerging approach to architecture in which building performance is a guiding design principle. This architecture places broadly defined performance above, or on a par with, form-making; it utilizes digital technologies of quantitative and qualitative performance-based simulation to offer a comprehensive new approach to the design of the built environment.

In this new information- and simulation-driven design context, the paradigm of performance-based design can be approached very broadly — its meaning spans multiple realms, from spatial, social and cultural to purely technical (structural, thermal, acoustical, etc.). The increasing emphasis on building performance — from the cultural and social context to building physics — is influencing building design, its processes and practices, by blurring the distinctions between *geometry* and *analysis*, between *appearance* and *performance*. By integrating the design and analysis of buildings around digital technologies of modeling and simulation, the architect's and engineer's roles are increasingly being integrated into a relatively seamless digital collaborative enterprise from the earliest, conceptual stages of design.

The contents of this book emerged out of the symposium on "Performative Architecture" held at the University of Pennsylvania in October 2003. That event brought together some of the leading individuals from different realms — architects, engineers, theoreticians, technologists — who comprise the contributors to this book, with the aim of providing informed views of what is meant by performances *in* architecture and *of* architecture.

The idea for the symposium was based on discussions between Ali Malkawi and myself about the apparent "disconnect" between geometry and analysis in the currently available digital tools. Our initial scrutiny of the tools (the instruments) that are used to digitally simulate the performance of buildings — the tools that are increasingly accessible to architects and their consultants — provoked broader questions, such as to what extent performance actually influences design and what performance means in architecture.

As we engaged the theme in its broader dimensions, we discovered that little has been written about performance in architecture. Yet this term — performance — has been widely used by owners, designers, engineers, cultural theorists, etc. Performance in architecture increasingly matters; however, it means different things to different people.

The chapters in this book provide a diverse set of ideas as to how building performance is relevant today and how it might be relevant tomorrow for architectural and engineering design practices. The projects discussed provide snapshots of different approaches, grounded in actual practices already taking place.

As readers will notice, the meanings of performance in architecture are indeed multiple and intertwined, and are irreducible to a simple, succinct definition. Performance, however, will increasingly underlie discussions about architecture in the future.

1

ARCHITECTURE'S UNSCRIPTED PERFORMANCE

DAVID LEATHERBARROW

1

ARCHITECTURE'S UNSCRIPTED PERFORMANCE

DAVID LEATHERBARROW

1.1
Neurosciences Institute, La Jolla, California (1992–95), architects Tod Williams and Billie Tsien.

The world is not an object such that I have in my possession the law of its making.

Merleau-Ponty[1]

I think that in every building, every street, there is something that creates an event, and whatever creates an event, is unintelligible.

Jean Baudrillard[2]

This chapter argues for a shift of orientation in architectural theory and practice, from what the building *is* to what it *does*, defining the first by means of the second. Broadly speaking, there are two ways designers and critics tend to view buildings: 1) as objects that *result* from design and construction techniques; and 2) objects that *represent* various practices and ideas. Although these accounts seem to explain fully the building's origin and destination, technological and aesthetic styles of thought reduce architecture to our concepts of it. Other and essential aspects of buildings come into view if one supposes that the actuality of the building consists largely in its acts, its *performances.*

The aim of this chapter is to outline how the building discloses itself through its operations (figure 1.1). For this to be apparent, constructive and perceptual intentionalities must be temporarily put out of play, not because they are misleading or wrong, but because they are normally taken to be fully explanatory. To see how the building itself operates, technological and aesthetic explanations must be temporarily suspended. This means subordinating, at least for a while, the questions about experience, meaning and production that normally occupy our attention. The autonomy thereby granted the building is contingent on this methodological premise. At risk in such an approach is architecture's perfect rationality, for it will be seen that performances or events depend in part on conditions that cannot be rationalized. This does not mean they cannot be understood, just that they must be understood differently.

Before proceeding, a certain assumption about architectural performance needs to be rejected; namely, that the development of new instruments and methods of predicting the building's structural or environmental behavior will radically redefine the discipline's practice and theory. Perhaps attention to performance will contribute to a new understanding of the ways buildings are imagined, made and experienced. But this new understanding will not result from the development and deployment of new techniques alone. The continued dedication to a technical interpretation of performance will lead to nothing more than an uncritical reaffirmation of old-style functionalist thinking — a kind of thinking that is both reductive and inadequate because it recognizes only what it can predict. I will return to this point below.

1.2
Philadelphia,
Pennsylvania.

ARCHITECTURAL PERFORMANCE

What is meant by architectural performance? The term is not new in architectural discourse, but its current use draws upon non-architectural linguistic traditions. Is the performance envisaged for the building like that of a machine or engine — the car on the street or the stereo in the study — or, is it closer to what might be seen on a theatrical stage or heard in a concert hall? This inquiry's central question can be stated simply: in what ways does the building act? What, in other words, does the architectural work actually do?

One way of resolving these questions rather quickly is to say that the building acts to "house" activities and experiences; an auditorium, for example, houses lectures, likewise a kitchen cooking, a courtroom trials, and so on. While this answer contains a germ of truth, the life of buildings that is predicated on use can be characterized as a *borrowed* existence, for it assumes that the room's or street's recognizable profile conforms to our expectations of it.[3] For the premise of predicated meaning to be sustained, the side of the subject must be taken. The significance that buildings possess is granted to them by you and me. Now, what seems eminently sensible from a pragmatic and pedestrian point of view has been shown to be naïve in Aldo Rossi's critique of functionalism.[4] Uses, he points out, often change throughout the life of the building — private houses become clinics, theaters apartment blocks, and so on. A criterion so inconstant as functional use cannot, he suggests, be used to define the building itself. Decisive instead, for Rossi, is type. For me it is operation or performance.

So a basic question presents itself: must the side of the subject be taken when an account of specifically architectural modes of behavior is given? Might this not leave something out, perhaps something essential? Certainly buildings are designed and built "for us": a farmhouse for farm life, a schoolhouse for schooling, and so on. A definition derived from Aristotle — architecture imitates human action and life — may be ancient, but it is still largely true. Granting this, can the building not also be understood apart from us and use, irrespective of programmatic requirements, individual desires, and cultural expectations? If not fully, can it be understood at least in part, without turning to ourselves as the benefactors of its identity? If we slacken the threads of intention that bind us to objects, what will appear?

The question seems worth pursuing because it is undeniable that rooms predate our use of them. They also remain as they were once we have finished with them. With just this single and simple observation about the building's extended temporality in mind, can it not be said that architecture exists quite happily and completely without us, that it is not entirely determined by "anthropological predicates" but is articulate on its own terms, that it is to some degree un-predicated, even auto-predicated? The turn to "experience" in architectural discourse, often announced with all good intentions, is generally a secret turn to design and production, insofar as the perceived is taken to be what is offered in designed perspectives. While congenial to technical or professional interests, this turn might well cause us to miss the reality of the building itself — especially that architectural reality that stands there irrespective of the vagaries of my interests or yours. My working hypothesis is that the theme of performance is a key to the building's *internal* definition or pre-predicated existence.

1.3
Experience Music Project, Seattle, Washington (1997–2000), architect Frank O. Gehry and Associates.

1.4
The façade detail, Phoenix Central Library, Phoenix, Arizona (1989–95), architect Will Bruder with Wendell Burnette.

1.5
Phoenix Central Library: The view from within.

Again, there are two common ways of missing the reality of the architectural work: one is to see the building as nothing but a system of components intended in design and realized by construction, the other is to view it as a system of representations outlined in composition and experienced in perception. Both make the building into an object, the first a result of technical reason and the second a confirmation of aesthetic expectations. This is precisely what happened in modernist functionalism, especially in the version advanced in the post-World War II period, after modernism's narrow determinism had revealed its incapacity for providing a plausible and legible urban architecture (figure 1.2). The debates on monumentality, for example, clearly testify to a widespread recognition of the poverty of functionalist solutions in their call for new alliances with art practices in the formation of urban centers. José Luis Sert's theory[5] and work exemplify this stance very well, as they demonstrate the desire to couple technical and aesthetic concerns in the formation of new civic institutions. Marcel Breuer's writings[6] and buildings demonstrate a similar thesis. Mid-century writings that address technical problems also assert the limited significance of functional concerns (the writings of Olgay and Olgay[7] are a good case in point, also the work of Max Fry and Jane Drew,[8] for both pairs suggested that art could compensate for the cultural sterility of functionally determined solutions). The widely celebrated buildings of our time no longer insert art into functional solutions, but they use it to drape or cover them: yet here, too, sculptural form is essentially a compensation for the inadequacy of functionalist solutions (figure 1.3).

Rather than rehearse this old debate between works that are useful and beautiful, seeking new answers to questions that were poorly formulated in the first place, it may be helpful to ask not about the work but about the way the work *works* (figures 1.4 and 1.5). Is there "action" in architecture's apparent passivity, in its steady and static permanence? Is the application of the term "behavior" to architectural elements anything more than a pathetic fallacy, or do buildings perform in some way? Compared to dance and musical expression, the building seems to be resolutely — even embarrassingly — inert and inactive; *frozen* music indeed. Compared to film, architecture seems positively motionless, about as animated as a stop sign. The house, theater and museum just sit where they have been planted, patiently awaiting a visitor's arrival and experience, as if they could only glow with life when ignited by interests you and I bring into them when we walk through their front doors. But is the building only what we make of it? One suspects there must be something more to it because if it were only the consequence of an inhabitant's intentions, it would be impossible to understand why we often feel the need to habituate ourselves to buildings, and also why they can alternately depress and delight us.

EFFECTS, ACTIONS, AND EVENTS

No doubt the building is a technical and aesthetic work, but it is known as such through its workings, through its instruments and equipment (broadly conceived). Stated more strongly, the building *is* its effects, and is known primarily through them, through its actions or performances. What is true for people is also true for buildings — character shows itself in what they do, in the decisions, choices or actions they take. The real locus and realization of character is in action — not in signs of identity added onto surfaces. The difficult point is that these workings do not arise from capacities and conditions produced by technical reason or aesthetic intentionality alone. One could assume these operations intend the building to work "for us," but, insofar as we will be followed by others who will have different expectations, this must be a very generalized "us." The generality of this determination is so great that it is insufficient to understand the building's full actuality. To grasp that, other kinds of performance must be described — not performances *in* architecture, instead performances *of* architecture.

For the sake of an example, imagine a lecture theater. Certainly such a setting can be described objectively: the four walls, the false ceiling that conceals air handling equipment, the sloped floor, the seats arranged for the auditors, the lectern for the speaker, the gathering space on the other side of the doors that benefits from natural lighting, and so on. All of it exists in indisputable factuality, all of it stands within the building on the site in the city with impressive permanence and stability, waiting for events with unequaled patience. But is this availability, this "permanence in waiting" objective or object-like in the way we commonly assume? Can such a room's performance be measured like other objects in the world, will it yield itself to techniques of sizing, weighing and computing? Is the room a machine for living in?

The inadequacy of this conception can be seen if the room's performance over time is considered.

Consider first the room's pre-history. Insofar as the lecture hall we have imagined has been arranged for the uses we intend, it existed before we walked though its doors. Because of this, it imposed itself on experience. Taking an abstract view of the situation, one can say some of its characteristics could have been foreseen. But in its concrete actuality every room is encountered as something donated to us from a past into which we have no real insight, and over which we have absolutely no control. Of rooms we commonly say they are *given*, but we generally have no knowledge of the people responsible for the gift, or of their specific intentions, desires and expectations for its use. Once seen, the room might be judged marvelous, singularly depressing, or largely insignificant because it is so typical; but until it is seen in its concrete actuality, it is unknown and (in its specific qualities) unexpected. Whether great or small, there is always some surprise when buildings are entered, and this results from the fact that the particularity of each comes from a past of which we are largely uninformed, each is "charged with a history that exceeds memory,"[9] the result of unknown and unknowable initiatives.

In truth, we do not so much enter rooms, but rooms (so to speak) happen to us. One way to begin thinking about what may be called the *event-character* of a setting is to consider its emergence out of a causality that no one understands very fully. Here we touch on an aspect of events that is essential — their unknowable beginnings and unexpected occurrence. When we use the expression "that was some event," we acknowledge the unexpected quality of what occurred. We give such an experience the name *event* precisely because of the unforeseen character of what happened — real events are always more than what we expected of them. Operations can indeed be managed, functions can likewise be scripted, but the events we take as important *cannot* — or what is planned is not what makes them important. Similarly, events do not result from technique or technical knowledge. This is because *foresight* is essential to technological thought.[10] Technique is always anticipatory, it is a form of knowledge that leads to preconceived results. Because events arise out of a past that we do not know, they cannot be produced technically. Putting the matter more forcefully, performative architecture is not the outcome of building or design technology, even up-to-the-minute digital technology. All that technique can give architecture is enhanced functionality.

The room's "*eventmental*"[11] nature is especially clear when its present appearance is considered. Viewing the way an auditorium presents itself on a given occasion, one could, speaking in generalities or broad abstractions, imagine the ways it subsists in its "indifferent emptiness" between various events. If the room's actuality is important, however, a wiser procedure would be to observe and describe its particularity on a given afternoon, filled with a particular event, on an announced topic, and so on. The setting stages an event, the success of which *at any given moment* is unknown. Is the room working? Are its provisions serving the aims of the speakers and audience? One could certainly say the room is generally adequate, but not until the event unfolds — in its unparalleled particularity — will anyone be able to say whether or not the room and the event was a success or failure. Both the singularity of the occasion — the working of the room for a specific occasion — and its "*undecidability*" in the present must be stressed. Even if inheritances confer orientation, they offer no guarantees. As with its past, the room's present condition is unknowable, but also unrepeatable, and cannot be constituted as such by any explicit intention. The event, as we say, is or is not happening. No single contribution, no matter how well-planned or thought out, can control its unfolding, not the schedule of the presentations, the refreshments for the breaks, the equipment provided, nor the soundproofing of the room will allow one to fully anticipate the outcome. If the event is only what was anticipated, it will have been both uneventful and unmemorable. Viewing the way an event unfolds in the present, we can discern an essential aspect of settings: place-bound events that truly merit the name arise out of themselves, despite my interests or yours, as if they were indifferent to them.

Lastly, in this little sketch of the auditorium's temporality, its future requires attention. How will a given event play itself out, a conference let's say? How will the room perform for a later speaker in the afternoon when the time after lunch brings a little sluggishness? Can one say a listener's inattention during a later speaker's talk will have resulted from the quality of the argument, the heaviness of lunch, or the way the room's atmosphere blankets a person's awareness? And what about one's understanding of the talks themselves, the differences between what the speakers will want to say and what each of us will hear, about all the disagreements, distortions, silent understandings, exaggerations and so on? Can any of that be known objectively? In order to provide an adequate account of what will have occurred one would need an unending hermeneutic. The event in, and of, a particular room is a phenomenon without a clearly known or knowable boundary, end or identity. What objects possess in abundance — sharp definition — events lack almost entirely. Put differently, the performance of a setting can only be known on its own terms, or, as I suggested earlier, *pre-predicatively*. Events cannot be defined, organized or

1.6
Aurora Place, Sydney, Australia (1996–2000), architect Renzo Piano Building Workshop.

1.7
L'Institut du Monde Arabe, Paris, France (1981–87), architect Ateliers Jean Nouvel.

scripted because their beginning, middle and end resist objective comprehension. This leads to a first conclusion: to understand architecture's performative character we cannot rely on transparent and objective description alone, or on techniques of quantification and measurement.

THE DEVICE PARADIGM

In what types of places do architectural operations unfold? Where are architecture's unscripted performances typically staged? One obvious answer is the building's moving or (more exactly) moveable mechanisms. This is an undeniable kind of "action" in architecture, for some of the building's parts literally move or allow themselves to be moved: apertures, screens, furnishings, etc., each of which has its own "range of motion," its stops, levels, intervals, etc. that anticipate and regulate its shifts and repositionings. Movements of this kind are variously manual and mechanical, initiated by human or environmental prompts, and controlled by manual, electrical or digital mechanisms. Their tasks, in general, are the modification and mediation of the environment in its widest sense, from climate to human behavior. The work of Renzo Piano's Aurora Place in Sydney, Australia (1996–2000), for example, could be used to illustrate this aspect of architectural performance. Each of its exterior surfaces, and all their elements, consist of moving and moveable mechanisms (figure 1.6). Perhaps the most famous early twentieth-century example of a building that presents itself as an ensemble of adjustable equipment is the Maison d'Alsace (Paris, 1928–31) by Pierre Chareau, with its exceedingly elaborate apparatus of ladders, screens, shades and so on. Roughly contemporary, and offering equivalent devices on the building's exterior, is Giuseppe Terragni's Casa del Fascio (Como, Italy, 1932–36). Perhaps the grandest example from these same years is Le Corbusier's Palace of the Soviets (1931) — huge sections of the building were meant to move with the actions of the assembled multitude.

Design of this sort follows what might be called the *device paradigm*.[12] The positions each element can take — the stops, levels and intervals — script the device's performance. Typically, these positions outline or frame a range of movements, normally from open to closed (figure 1.7). The intelligence of a device is measured not by the

1.8
The upper deck,
Neurosciences Institute,
La Jolla, California
(1992–95), architects Tod
Williams and Billie Tsien.

1.9
Neurosciences Institute:
the central court.

breadth of this range, nor by the number of intermediate positions, but by its capacity to adjust itself to foreseen and unforeseen conditions. An analogy that may be useful here is with musical or theatrical improvisation, as if the stops and positions of the building's elements do nothing more than sketch out the guidelines of a performance, allowing for spontaneous qualifications that attune the ensemble to particular conditions, as they vary over time. Approximate movements can be intended, but settings can also yield, respond or react to unforeseen events. The architectural drama, then, comes alive through the building's performances. The first step in the development of a performative architecture is to outline strategies of *adjustment*.

ECONOMY OF PERFORMANCE

There is another site of architectural action in which performance is less obvious but no less determining: those parts of the building that give it its apparently static equilibrium, its structural, thermal, material stability. When discussing these elements (columns and beams, retaining walls and foundations, but also cladding and roofing systems), it is common to talk of their "behavior" — not only talk of it but to anticipate it, even predict it. Obviously, talk of this sort is metaphorical, but in truth the building must work at staying as it is. It must *work with* ambient conditions, such as gravity, winds, sunlight and so on. It must also *work against* these forces. And it must suffer their effects. No actor on stage ever suffered as much as buildings do — whether one thinks of use and misuse, weathering, or additions and alterations.

The economy of performance — in a site, as if on a stage — is always an exchange between forces and counterforces. To act is to counteract. The building's labor is quite simply the amount of effort it takes to sustain this economy, to keep up or play its part. The term we use most frequently for this work is *resistance*. The work of Tod Williams and Billie Tsien, especially the Neurosciences Institute in La Jolla, California (figures 1.1, 1.8 and 1.9; 1992–1995), from a few years ago demonstrates awareness of this kind of performance — working with and against its site — also, much more locally, against the

1.10
Margaret Esherick House, Philadelphia, Pennsylvania (1959–61), architect Louis I. Kahn.

1.11
Ise shrine, Ise, Japan, last ritually reconstructed in 1993.

pressures of human touch, evident in its variously rough and smooth surfaces. A similar, and well-known, case from the twentieth century is Mies van der Rohe's Tugendhat House (Brno, Czechoslovakia, 1928–30). Less well-known, but even more vividly engaged with its site, is his Wolf House (Guben, Germany, 1925–26).

Speaking very generally again, the work the building performs involves resistance, by means of which its capacities and identity become apparent. The façade, in fact, is the site of precisely this resistance, offered by the latent qualities of materials against ambient forces (figure 1.10). Should one say, as I did earlier, that the building's destiny is to suffer? Is its work passive? That depends entirely on what is meant by the term. Is it fair to say that the sprinter poised for the start of a race is passive? Is it not more accurate to observe that the explosion ignited by the starter's pistol presupposes a coiled potential that can only be constituted and maintained through strenuous effort? Is the building's action against the steady pressure of the hillside into which it is cut any different? In both there is force and counterforce, which suggests an inevitable contextuality of the building's performative elements, by which its equipment transcends itself into a range of spaces and regions in the same way that it transcends itself into several temporalities — disavowing, again, its status as an object or phenomenon that can be objectively defined.

The design for performance of this sort is based not on a device but on a *topography paradigm*. Movement here is not the change of position but of state. The force–counterforce relationship results in alterations to the building's physical body that demonstrate its capacity to respond to ambient conditions. Stains on the building are evidence of its capacity for resistance. Cracks in the wall show limited success on this front.

The obverse of cracking and staining is shaping and finishing — whether that of construction technology, environmental influence or everyday use. This kind of suffering (architectural *pathos*) was described by Peter Zumthor as enrichment.[13] Buildings, he said, take on a beautiful and specific richness when traces of life are sedimented onto their surfaces (figure 1.11). Movement, action or performance in the so-called static permanence of

1.12
The Getty Center, Los Angeles, California (1984–97), architect Richard Meier and Partners.

buildings or elements is toward or away from the fullness of their potential — movement typically described as development or deterioration. Because not one of these kinds of movement can be precisely predicted, architectural performance such as this can be described as unscripted.

Operations in and outside the building are dependent on several contingencies: those of the inhabitant's interests and vantage, of the climate, the seasons, and of the times. The building's workings are also dependent on changes to those parts that have been joined together to form the work. No doubt it is obvious to state that over time the building's materials eventually fail; but we rarely think seriously enough about this inevitable or essential indeterminacy. Materials suffer and vary at different rates. Some parts of the structure settle and move, others not, or not much, while still others suffer a range of surface variations. In the face of these alterations, maintaining equilibrium among the building's parts is a task that cannot, in principle, be completed. Nor can its difficulties be foreseen. In the unfolding of the operations that sustain a dwelling situation, architectural elements constitute themselves into something of a stable ensemble that possesses a comprehensible but provisional finality, for such a configuration is always and only temporary.

Concerning the changes the building suffers as a result of "external" contingencies, there is some degree of predictability of developments, resulting from past experiences, but never certainty. Some locations show greater constancy of climatic conditions, or less seasonal change — such as the Caribbean — while others, like Canada, show continual alteration. In Montreal, they say if you do not like the weather, wait ten minutes. But even in moderate zones, the environment sometimes acts in ways it is not supposed to, as the history of tropical disasters proves. If the ambient environment was steadier in its offerings, the building could assure itself of the adequacy of its provisioning and would not need to continually adjust itself. In these circumstances (which are really those of the laboratory), performances could be scripted. But the world in which buildings actually exist is hardly so lawful.

The true measure of a building's preparations is their capacity to respond to both foreseen and unforeseen developments. Stated in reverse, bad buildings are those that cannot respond to unexpected conditions because they have been so rigidly attuned to environmental norms. Our tendency to think of environmental conditions as external or extrinsic contingencies should be resisted, for no building operates without them or apart from them. Jean Nouvel claims to put nature to work in his architecture, for the beauty and richness of his surfaces do not result from design or construction technique alone but also from the action of ambient lighting, which variously and wonderfully saturates the skins with fluctuating qualities — transparency, opacity, reflectivity and color. The environment — in this case, natural light — must therefore be seen as internal to the building. Workings of this sort are evidence of engagement between what was and what was not constructed, of the building's willingness (or need) to interact with what it is not. Here again, non-objectivity, as contingency, enters the heart of the work, through its operations. Here, also, emerges the possibility of a form of representation internal to architecture; for when the building *identifies itself* with its milieu it becomes something it is not, the way all representational figures do. Put differently, the building's performance is the key to its natural (unimposed) symbolism, because when the building defines itself in terms of what it is not (the natural and cultural milieu) it inaugurates precisely the sort of self-negation that is necessary for representation to occur (figure 1.12).

When the building is freed from technological and aesthetic intentionalities, we discover its lateral connections to an environmental and social milieu that is not of anyone's making, still less of design and planning. And it is precisely these connections that animate its performativity, even if they cause the building's work to resist both conceptual mastery and exhaustive description. The point to be stressed is the building's eccentricity, its existence outside of itself, for its behavior testifies to a constitutional weakness at its center, a negativity at its heart, because it must wait on the environment to give it what it lacks — light, air, human events and so on. Still, what the environment offers is always somewhat different from what was expected. The building's internal disequilibrium obliges it to accept into its make up conditions over which it has no control.

With the different dimensions of the building's contingency in mind, a second conclusion can be proposed: that architecture's performative labor has no end, for it is a task that continually presents itself anew.

TOPOGRAPHIES OF PERFORMANCE

Performance in architecture unfolds within a milieu that is not of the building's making. A name for this milieu is *topography*, indicating neither the built nor the un-built world, but both.[14] Three characteristics of topography sustain the building's performativity: its wide extensity, its mosaic heterogeneity and its capacity to disclose previously latent potentials. There is always more to topography than what might be viewed at any given moment. Excess is implied in its ambience, for what constitutes the margins of perceptual concentration always exceeds the expectations of that focus. But this still more of topography, this outward increase of breadth and compass, does not offer to experience more of what is locally apparent. Differences are always discovered in the spread of topography; contrast and complementarity structure relationships between its several situations and sites.

In modernist theory, space was presented as the all-embracing framework of every particular circumstance, the unlimited container of all possible contents. Likewise in modern science, continuous space was understood to be isotropic and homogenous, possessing a self-sameness congenial to intellectual mastery because of the conceptual character of its attributes. The topography in which buildings perform is just the opposite of space: polytropic, heterogenous and concrete; its regions contrast, conflict and sometimes converse with one another. Yet it is not a field of infinite difference either, for it continually offers experience of both unexpected and familiar situations. If space advances its array all at once (in simultaneity), actual topography gives its locations through time. In any given site, at any given moment, its structure requires that some places be recalled, others anticipated.

1.13
Santa Fe Art Institute, Santa Fe, New Mexico (1996–99), architect Legorreta + Legorreta.

Topography's latency is apparent if one considers the way it gives itself to experience. Like events, landscapes — whether they are urban or not — contain unforeseen potentials, and show these potentials in the various ways they offer themselves to perception. The word "capacity" applied to physical things indicates similarly unseen possibilities. Capacities cannot be (fully) discerned because they keep themselves recessive — like the backside of an object one is looking at or the inside of something shaped or polished. Construction finishing aims to cultivate the potential of things. Because it too is material, and can be cultivated, topography is *not* what physically appears in a given place — built or un-built — or *not only* that. Sites are surveyed in the early stages of design so that given conditions can be described and understood. While this seems obvious, the term *given conditions* is far from clear. We tend to assume that the place exhibits "its intentions" the way designs present theirs; in both, intentions are *shown*, and "givenness" we believe offers expressive display. But this again confuses the standing of a figure with that of its ground, for the topography in which architecture performs is not composed of objects in the same sense; it does not expose the grounds (intentionality) of its formation, but serves as the grounds for that formation (figure 1.13).

If topography's potentials exceed one's grasp and remain unforeseen to some degree, they can also be said to be unreasonable, at least in some measure. If, as argued previously, the building is always or necessarily engaged with topography, by virtue of its inevitable contextuality and contingency, its performances, too, will be (to some degree) unplanned, or they will arise from "causes" that are unassignable.

This suggests a different understanding of the building: it is not a technical preparation or not that chiefly, nor is it primarily a representation of such a preparation, but it is a non-technical and non-aesthetic performance, the designer's comprehension of which acknowledges its continual need for readjustments in order to reclaim its own equilibrium and to sustain its engagement with un-built or previously built contingencies. Put more simply, the building's approximate disequilibrium animates a life and a history of ever-new performances (figure 1.14).

Aristotle once advised that the mark of a wise individual is to strive for the degree of exactitude in descriptions that is appropriate for the given subject. The same exactness, he said, must not be sought in all departments of philosophy alike, any more than in all the products of the arts and crafts. Let me cite him: "It is the mark of an educated mind to expect that amount of exactness in kind which the nature of the particular subject admits."[15] For this reason it is equally unreasonable to

1.14
Carpenter Center for the Visual Arts, Cambridge, Massachusetts (1960–63), architect Le Corbusier.

accept merely probable conclusions from a mathematician and to demand strict demonstrations from an orator. In a similar vein Aristotle recommended that when building a house, sketches of basic (configurational) principles should be made in outline form only, so that they can be gradually filled in as unforeseen exigencies and opportunities arise. The carpenter and geometrician both seek after the right angle, he said, but in different ways: "the former is content with an approximation to it which satisfies the purpose of the work, the latter looks for the essence or essential attributes."

INSTRUMENTALITY PLUS

At the outset I distinguished between two kinds of understanding in the theory of architectural performance: the kind that can be exact and unfailing in its prediction of outcomes, and the kind that anticipates what is likely, given the circumstantial contingencies of built work. The first sort is technical and productive, the second contextual and projective. There is no need to rank these two in a theory of architectural performance; important instead is grasping their reciprocity and their joint necessity. If acceptance of an uncertain foundation for performance seems to plunge practice into irrationalism, we need only remember that most of the decisions we make in our daily lives rest on a foundation that is just as uncertain. The cultural norms that serve as the horizon of unreflective existence will not stand up to rational scrutiny, but are not for that reason nonsense, nor are they opaque to understanding. They are certainly transparent enough to sustain debate, the result of which is adjustment or alteration. For a theory of performativity we should seek nothing more and nothing less: instrumental reason and the rationality on which it depends, *plus* situated understanding that discovers in the particulars of a place, people and purpose the unfounded conditions that actually prompt, animate and conclude a building's performances.

NOTES

1 Maurice Merleau-Ponty, "The Philosopher and His Shadow" in *Signs*, Evanston, IL: Northwestern University Press, 1964, p. 180.

2 Jean Baudrillard and Jean Nouvel, *The Singular Objects of Architecture*, Minneapolis, MN: University of Minnesota Press, 2002, p. 16.

3 The term "borrowed existence" is derived from Jean-Luc Marion's *In Excess: Studies of Saturated Phenomena*, New York: Fordham University Press, 2002. While my argument is indebted to his entire account of "givenness," his description of borrowed existence can be found in chapter 2, "The Event or the Happening Phenomenon." I have also adapted his interpretation of the event structure of the lecture theater.

4 Aldo Rossi, *The Architecture of the City*, Cambridge, MA: MIT Press, 1982, especially "Critique of Naïve Functionalism," and "How Urban Elements Become Defined."

5 See, for example, Sert's contribution to the CIAM 8 conference: "Centres of Communal Life," in J. Tyrwhitt, J. L. Sert and E. N. Rogers (eds), *CIAM 8 The Heart of the City: Towards the Humanisation of Urban Life*, New York: Pellegrini and Cudahy, 1952, pp. 3–16. While his work is also documented in this publication, see more fully Knud Bastlund, *José Luis Sert Architecture, City Planning, Urban Design*, New York: Frederick A. Praeger, 1967.

6 See Peter Blake (ed.), *Marcel Breuer: Sun and Shadow, The Philosophy of an Architect*, New York: Dodd, Mead & Company, 1956.

7 Most helpful among their writings on this particular topic is Aldar and Victor Olgay, *Solar Control and Shading Devices*, Princeton, NJ: Princeton University Press, 1957.

8 See, for example, Maxwell Fry and Jane Drew, *Tropical Architecture in the Dry and Humid Zones*, Malabar, FL: Robert E. Krieger, 1964.

9 *Op.cit.* Marion, *In Excess*, p. 34.

10 This point is elaborated in David Leatherbarrow and Mohsen Mostafavi, postscript to *Surface Architecture*, Cambridge, MA: MIT Press, 2002.

11 This term is Jean-Luc Marion's. See *In Excess*, p. 32.

12 This term, used more expansively, can be found in Albert Borgmann, *Technology and the Character of Contemporary Life*, Chicago, IL: University of Chicago Press, 1984, especially see chapter 9.

13 Peter Zumthor, "A Way of Looking at Things," 1988, in *Thinking Architecture*, Baden, Switzerland: Lars Müller, 1998, p. 24.

14 For more on topography in this sense see David Leatherbarrow, *Uncommon Ground: Topography, Technology, and Architecture*, Cambridge, MA: MIT Press, 2002, and David Leatherbarrow, *Topographical Stories: Studies in Landscape and Architecture*, Philadelphia, PA: University of Pennsylvania Press, 2004, especially the conclusion, "Ethics of the Dust," which elaborates the points summarized here.

15 Aristotle's text, *Nichomachean Ethics*, 1.3.1, is cited and discussed in Wesley Trimpi, *Muses of One Mind*, Princeton: Princeton University Press, 1983, p. 125.

2

PRODUCT AND PROCESS: PERFORMANCE-BASED ARCHITECTURE

ANDREW WHALLEY

2

PRODUCT AND PROCESS: PERFORMANCE-BASED ARCHITECTURE

ANDREW WHALLEY

The evolution of late twentieth-century design can be investigated by considering some of its dominant themes, such as mass production, information technology, transportation and the workplace. It is already evident that new concerns will influence the future of architectural and industrial design. If any theme can characterize the new era, it is the changing perception of space and design's new fluidity.

Architecture and industrial design are both a response to and reflection of the society that we live in. Architects produce designs by carefully analyzing requirements and creating thoughtful solutions. A combination of intellectual and manufacturing capabilities enables them to do this — essentially, to generate and develop the idea, the process and the product.

Grimshaw is a process-driven practice. There is no preordained stylistic solution but rather a rigorous exploration of ideas resulting in a strong and legible concept — a design diagram. The solutions evolve out of a broad investigation and understanding of a project's program — the careful balancing of elements that make-up architecture. The investigation continues down to the finest detail, so that the overall concept is evident in the smallest part of the resulting building.

Grimshaw is also a product-driven practice that takes a very pragmatic approach to architecture, which has evolved from an understanding of manufacturing and an appreciation of the way things go together. This initially may appear to be an overtly functional way of looking at things, but we believe that out of that functionality, understood in performative terms, beauty arises.

ARCHITECTURE OF CHANGE

To understand the firm's design philosophy and its consistent application over the past few decades, it is important to locate the practice's work within the context of time and technological development.

The initial work at Grimshaw, in the 1970s, was founded in industrial architecture, of which an early example is the factory in Bath, UK, for Herman Miller, the furniture manufacturer (1976). The client wanted a factory with the potential for change, because it did not know what lines it would bring out over the next twenty or thirty years. The solution was an early use of fiberglass paneling to provide a very flexible, adaptable skin (figure 2.1). Along with this adaptive cladding system on the outside was a very flexible servicing strategy inside. The combination allowed the building to behave like an organism that can adapt to suit different demands. Over fifteen years of use, the factory has been rearranged five times. In the most recent change, the occupants moved the canteen to an area that was formerly used for manufacturing and so it had an opaque skin. By moving glazed panels to that area, views onto the river were opened up. By embedding adaptability into the design of the building and its systems, we created an architecture that not only performs over time but that also improves the quality of life for its users.

For the second Herman Miller project at Chippenham (1982), also in the south west of England, Grimshaw again wanted a building that could anticipate change. It had to have a large span and be essentially a big warehouse, but one with the scope for conversion to office use in the future. Long-term flexibility had to be anticipated. We designed a whole cladding system (figure 2.2) that would meet those performance criteria, because a warehouse is essentially about a skin. An office, however, is something more sophisticated — you have to have windows, doors and natural ventilation. The solution was to try to design a skin that would foresee that change and that could be adapted to suit its use.

It soon became apparent that "off-the-shelf" components from cladding manufacturers were too expensive, or did not even exist. For the Herman Miller warehouse, Grimshaw designed its own cladding system of pressed aluminum — one which we have continued to develop and improve over the years. To understand how elements such as cladding are made means going to the factories and exploring the material and manufacturing possibilities, which in turn informs the way architects detail and design a skin. Through this rigorous approach beauty can come from pragmatism.

2.1
Adaptive fiberglass paneling system, Herman Miller Factory (1976), Bath, UK, architect Grimshaw.

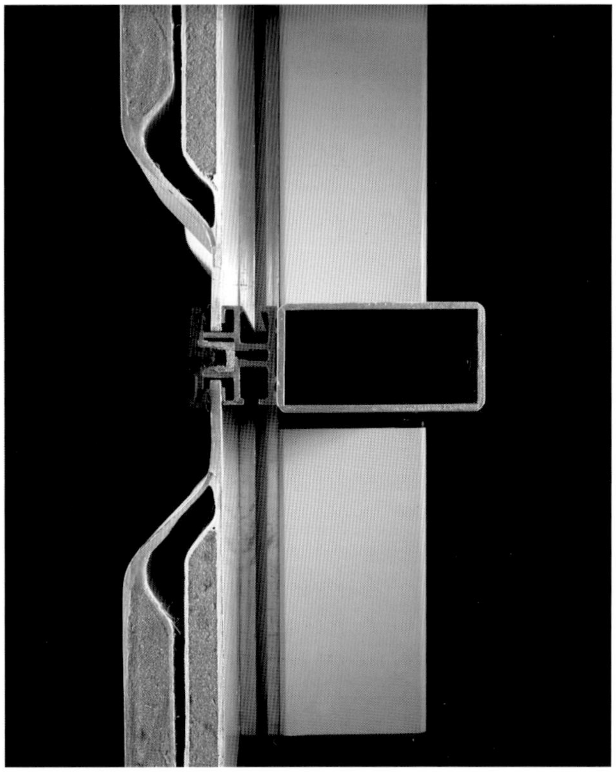

2.2
Adaptable cladding system, Herman Miller Warehouse (1982), Chippenham, UK, architect Grimshaw.

2.3
The façade detail, IGUS factory, Cologne (1990–2000), architect Grimshaw.

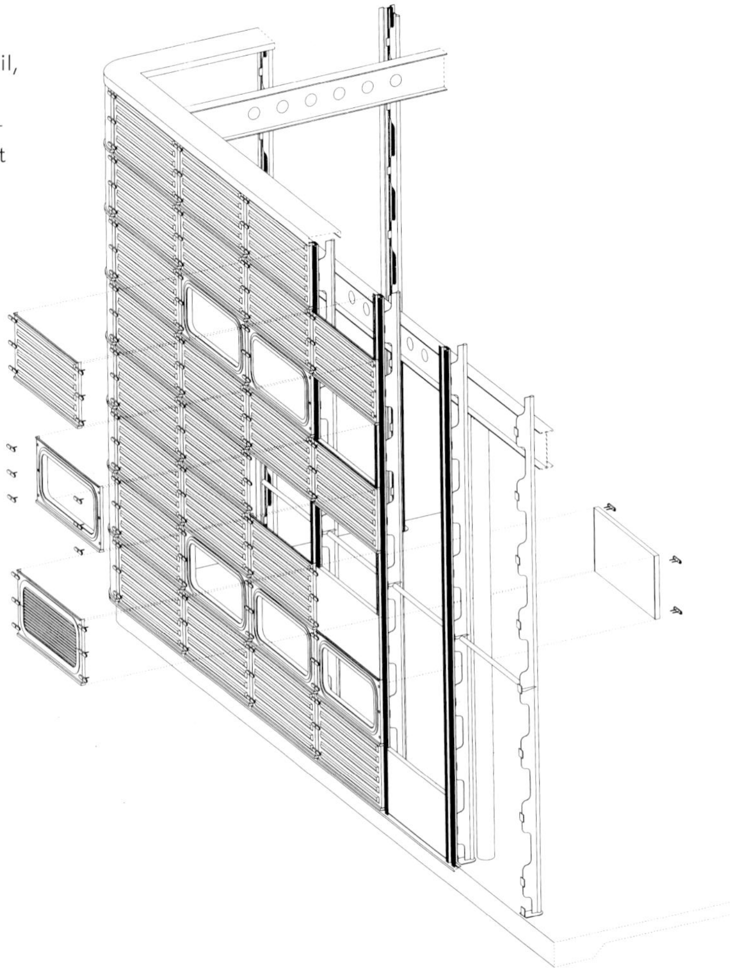

The practice pursued this approach in a number of buildings, such as the flexible factory designed for IGUS (Cologne, 1990–2000), where we have developed the vocabulary of the skin further with inspiration from Jean Prouve. The physically expressed clamps show how the skin works and how the panels are pressed back onto a framework (figure 2.3). That building is now up to phase five and has grown over the last ten years to about 400% of its original size, which has been possible because it was built around courtyards and based on a design that did anticipate change. The system of roof lights brings light deep into the building. A series of modular elements float inside the building, and can contain a variety of functions, such as office spaces or restrooms. The theater technology of pressurized air is used in the pad units under modules, so each can be moved manually around the factory floor. Rather like a chessboard, one can rearrange the entire building.[1] The units can access natural light and ventilation no matter where they are placed, by simply hooking up to one of the roof lights.

In the *Financial Times* building (London, 1993), the concept was all about the process of producing newspapers as part of a very dynamic, kinetic architecture. Commuters drove past the building at night as the newspaper was

2.4
The *Financial Times* Printing Plant, London (1993), architect Grimshaw.

2.5
Sketches and the prototype of the glazing bracket, The Channel Tunnel Terminal at Waterloo Station, London (1993), architect Grimshaw.

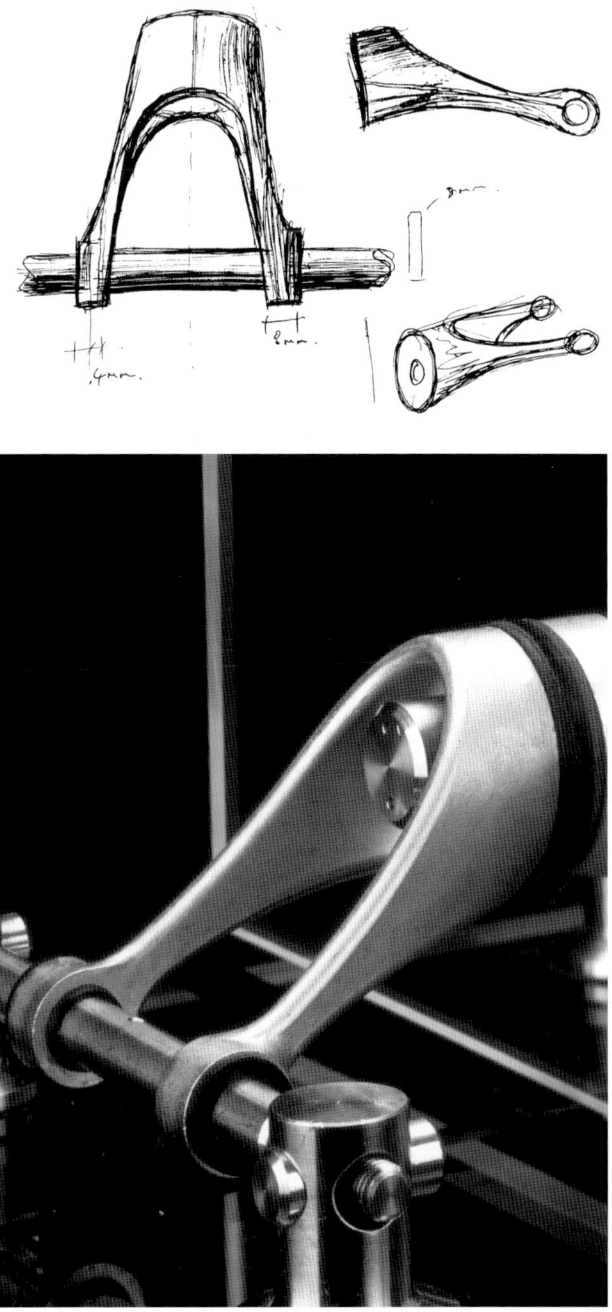

being printed and then read it the next morning. To show the kinetic infrastructure, we designed a vast glass wall (figure 2.4), placing the structure on the outside to articulate the building and to keep the inside unobstructed for the industrial process. But within ten years of its opening, the building had a change of use. The printing presses went out and it became an Internet switching center. Because of the robustness and flexibility in the architecture, the new owner made the change in about six months. So to some extent, one can build in flexibility as part of a "loose fit" approach.

DESIGNING THE BUILDING'S DNA

At Grimshaw, a key concept, such as the large glass wall at the *Financial Times* building, is followed to the finest detail. One can find the concept in the smallest detail that makes up the building and understand the larger picture from the finest elements. The same amount of time and rigor are given to all the elements, however large or small (figure 2.5). In designing a component of a glazing joint, for example, the architects make prototypes in the office in foam and wood.[2] These small elements can be seen as the DNA of our buildings.

2.6
The Channel Tunnel Terminal at Waterloo Station, London (1993), architect Grimshaw.

A very good example of that process, and one that shows the shift to a more dynamic architecture, is the Channel Tunnel Terminal at Waterloo Station (London, 1993). The brief called for a very large roof, sheltering a terminal that had to contain many of the program items that would typically be found in an airport rather than in a railway station (figure 2.6). Designed to handle 15 million passengers, it also had to be 'shoe-horned' into the center of the city. The railway engineers produced a complex footprint to allow for the trains coming into the station. The brief for the roof, which was only 10% of the capital cost of the whole project, had to enclose all that space, snaking its way to the terminus.

Again, a considerable amount of time was spent looking at manufacturing as ideas were developed. We made a series of models as a way of visualizing and understanding the space and structure, and then came up with an efficient asymmetrical form to suit the track layout. It encompassed the best of both worlds. While drawing inspiration from the great Victorian engineered railway sheds and triumphant halls of the nineteenth century, where the structure is expressed internally, it also created an unprecedented public façade. The three-pin arch form places the pin to one side, responding to the asymmetrical nature of the platform layout. The pin forms the point of contraflecture, which means that the compression and tensile elements reverse. We used this as a device to invert the relationship of interior to exterior structure. This exterior structure gave the building a public face to present to London, something a Victorian station never did. Materials were only used where they were

2.7
Waterloo Station:
the glazing detail.

2.8
British Pavilion at the World Expo '92, Seville (1992), architect Grimshaw.

needed structurally. The use of telescoping and tapering tubes produced a dynamic skeleton-like form, with opaque cladding facing the station and a glass elevation towards city, with views of the trains coming in and out.

The problem was, of course, how to create a glass envelope that could move and snake around the irregularly shaped site; for economy and expediency we wanted to keep all of the glass as a set of standardized rectilinear sheets. So we designed individual elements that would create the whole (which goes back to the DNA concept mentioned previously). The key element is a joint shown in figure 2.7. The lattice has lots of rectilinear sheets of glass similar to a Victorian greenhouse. The joint allows the different geometries to be used by letting each sheet slide, like the scales on a snake's skin. A considerable amount of time was spent designing a single component, a joint element, which could pick up the skin anywhere in space. This is how the design concept was fulfilled and the manufacturer satisfied. The overall effect is organic, fluid architecture, although all of the roof glazing is made out of rectilinear pieces of glass and tubular elements.[3]

While Waterloo was still being designed, Grimshaw won a competition to design the British Pavilion at the World Expo '92 in Seville (1992). The aim was to create an environment suited to an Expo in a very hot climate while also demonstrating a high degree of sustainability. We designed an enclosure that tempered and controlled the environment (figure 2.8), and we placed within it a number of highly conditioned pavilions to provide flexibility of use because the exhibition content was still to be established.

2.9
Ludwig Erhard Haus, Berlin (1991–1998), architect Grimshaw.

2.10
Ludwig Erhard Haus: Street level, interior view.

The concept for this building was to use the sun to cool the building. The roof was covered with photovoltaic cells and, together with fabric sails, shielded the enclosure from the heat. The resulting generated energy would pump water and pour it over the glass, allowing in light but keeping the building cool. A reasonable thickness of water absorbs almost all the infrared (heat) components of light, while still allowing the rest of the visible spectrum into the building. This idea of functional performance was manipulated further to give the space exceptional qualities that warm or cool the senses. Sculptor William Pye took the idea and developed the project's sculptural elements. He turned the water into droplets, which visitors could hear falling down as they walked through the space — they actually heard, saw and understood what was cooling the building. At night, the water surface was lit and the visitors could still enjoy it for its intrinsic qualities. The idea of using water worked on a number of levels — not just functional and not just performance-related — because the space was also sensuous and sculptured.

DESIGNING THE CITY SPACES

Change is also relevant on an urban scale. The Grimshaw approach can be seen in the Ludwig Erhard Haus (Berlin, 1991–98), which contains the Berlin stock exchange and offices (figure 2.9). The problem with office buildings in a city is that they impose their own grid and their own form onto the city's streetscape. The main concept, and one which is being followed in other new projects, was to create city-type spaces at ground level. The technically driven tectonic solution was to hang the upper floors from a series of structural arches, an approach that opened up all sorts of possibilities. By hanging the floors, the maximum enclosure of office space could be created without infringing rights of light and view angles. The resulting "soft" form is entirely described by what the practice was allowed to build; realizing this potential meant that it wasn't necessary to design a tall building. Consequently, the user can do anything at street level because there is a clear spanning structure with no columns (figure 2.10). The elements that are in there now can be substituted with something else in the future, such as an internal plaza or a skating rink.

2.12
Paddington Station: disseminating information.

2.11a–b
Paddington Station, London (1997–99), architect Grimshaw.

Grimshaw also designed the elements that take the visitors from that "city space" up into the offices. With this building, the brief was to create an office that would take three or four different organizations in one space. In the long term, it was not obvious how the building would be used, so it had to be organized to encourage people to mix and meet. The practice refers to this as the "mixing valve approach" where the social aspect of work is opened up, which is essential, as it is the people using a space that give life to a building. A central staircase with broad platforms encouraged everyone to change levels using central circulation; people could also use a bridge in the heart of the building. The building allowed discourse and chance meeting, a social aspect of the office environment.

Similarly, Grimshaw's work at Paddington Station (London, 1997–99) creates a city space where technology helps people use the space. Our aim was to reinvigorate the Isambard Kingdom Brunel station of 1854, as it had become very congested and we had to think of a new way to make it work. A new building was built behind the main concourse (figure 2.11a), with new technological elements to ease circulation (figure 2.11b). With the help of the Grimshaw industrial design team, banks of plasma screens in several locations disseminated information to stop people congregating around one centralized departure board (figure 2.12). Introducing these performance-related elements within a 150-year old building instigated a new way of working, while respecting the historic fabric of the existing volume and space.

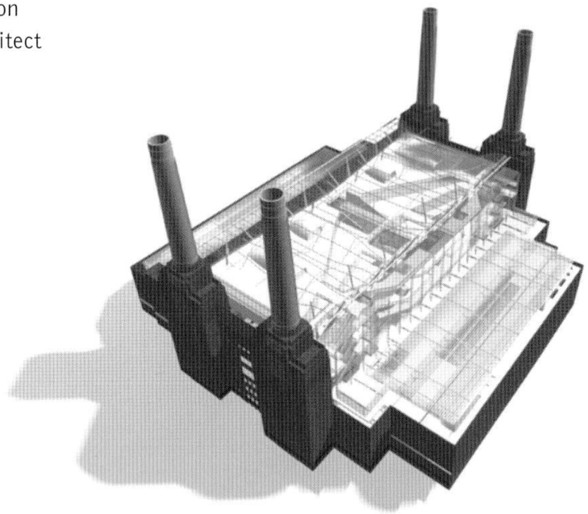

2.13
Redevelopment design for Battersea Power Station, London (2000–), architect Grimshaw.

PERFORMANCE-BASED DESIGN

One of the best examples of change in buildings in the Grimshaw portfolio is the redevelopment of Battersea Power Station (London, 2000–; figure 2.13). Designed to make power out of burning coal, Battersea Power Station is a massive, monolithic brick edifice that has been derelict for over thirty years, and has been partly demolished. Grimshaw has designed a set of components to create a lightweight roof for the old turbine halls. This solution — designed with the aid of non-linear mathematics — fuses twenty-first and early twentieth-century architecture to make a vast and flexible enclosure for this mixed-use scheme.

Grimshaw's brief for the Victoria & Albert Museum Boilerhouse Extension (London, 1996; figure 2.14) was to create a twenty-first century enclosure for digital art. We wanted to create a building that could showcase digital art and at the same time change, move, shift and replenish itself to reflect this ephemeral medium. A sculpture inspired the solution: a giant cube of ice by Anya Gallacio, in which the very beautiful platonic form slowly disintegrated over time and morphed into something quite different. Something crystalline and architectural, which disintegrates and changes, seemed to convey the idea of digital art and digital media. Again, technical solutions came into play. At that time we used a new type of glass (*Priva Lite* by Saint Gobain) that could change from clear to translucent (and vice versa), and which could be projected upon. When the glass is completely clear, one could see the functions and exhibits within the building. When translucent, it could become a medium for showing text, images or both. Like the ice sculpture, it could also change and seem to dematerialize.

We did not win the competition with that entry but, for Grimshaw, competitions are a chance to explore ideas in a relatively short space of time, as opposed to the often long gestation of a built project; in turn, these ideas can be developed and incorporated into other projects if they are not realized into buildings initially. In a successful competition bid for a new art gallery in northern Spain, Fundacion Caixa Galicia (La Coruna, 1998–), which is currently under construction, we revisited some of the ideas from the previous project and took them forward. The surrounding buildings have beautiful "gallerias," essentially double-skinned façades, which protect the buildings from the harsh environment of the Atlantic. A floating glass wall follows the line of these galleria façades, but the building then sweeps back physically to

2.14
Victoria & Albert Museum Boilerhouse Extension, London (1996), architect Grimshaw.

2.15
Fundacion Caixa Galicia, La Coruna, Spain (1998–), architect Grimshaw.

2.16
Donald Danforth Plant Science Center, Saint Louis (1998–2001), architect Grimshaw.

allow light down into a very deep basement (figure 2.15). The building is almost as deep as it is high, and is shaped and sculptured by its performance criteria, but at the same time it is sensitive to its setting. It was made using a new type of glass (*Holopro* by G+B pronova GMBH), that has a holographic-etched interlayer which can be seen or which can disappear. If an image is projected onto it, from the outside, it reads even in daylight as a sharp clear picture on a solid surface, but from inside the image is not seen and one can look straight through it as with clear glass. With no projection, it appears as a clear glass surface from both inside and out, and from the street one can see the building behind. So, by projecting onto the surface, the apparent position of the building envelope shifts and its relationship to the street changes — it is both a tectonic device and an urban mechanism.

At the Donald Danforth Plant Science Center (Saint Louis, 1998–2001), Grimshaw's first project built in the United States, the brief called for a series of laboratories. These pre-described, highly serviced and functional spaces usually come in standardized layouts. We did not challenge this, but instead focused on how the building is set within the site. Given that Saint Louis is quite hot in summer and very cold in winter, and that the circulation space was critical, we created a central social space which connected every part. In our early schemes we explored the concept of the skin having two different relationships, depending on the orientation. We designed a rigorously engineered façade relating to the south-facing main street (figure 2.16), protecting the building from the sunlight, but still offering limited views. From the other side, which is approached on foot, the building presents a more organic façade with a louvered, sawtooth-like skin. The resulting effect would have been that the building evokes a different response from a distance and from close at hand. From the road it is a precise, highly engineered building that one can look into to see the central circulation space and the library; one is given an entirely different experience when walking to the building from the adjacent parking lot.[4]

We have recently completed the initial design work for a new painting and drawing building for the Royal College of Art (London, 2000–). Again, the design is performance-driven to create the optimum painting and drawing conditions. There are very few purpose built, high-quality painting schools. The most famous is probably Charles Macintosh's Art School in Glasgow, Scotland, which has beautiful north-facing painting studios. The existing painting studios at the Royal College of Art receive direct

2.17
The painting and drawing building for the Royal College of Art, London (2000–), architect Grimshaw.

2.18
Royal College of Art: partial section.

sunlight, forcing students to cover up the windows with paper. The challenge was to work out how to create beautiful light and space on a very tight site. The answer was a skin that allows north light through but that shields the building from direct sunlight. This approach has resulted again in a building with a dual appearance: a fairly opaque building on the side facing the Albert Hall (a solid building that sits comfortably with the monumental buildings around it) and a very open glass building as seen from Kensington Gore (which is due north) (figure 2.17). The zinc cladding gives the building solidity from the south side, but then, as one moves around the building, it has a completely different appearance as it becomes more open.

At roof level, all the painting studios get the right quality and balance of north light; a high-quality light is also achieved for the studios on the lower levels (figure 2.18). As already mentioned, the building's design is performance driven. The roof concept and the way the artificial and natural light are mixed required considerable testing and the production of a series of different models. But, in addition to having a functional quality, there are sculptural and aesthetic qualities resulting from the way the sun moves across a very deep, highly modeled façade; the sun dances across it, animating the louvers and creating a very pleasing effect, which we simulated using computer models.

Grimshaw's work is evolving and this is partly due to the way computers are changing how architects think about space. An example is a small project for Cemex, a major Latin American construction company. They centralized all of their facilities on one site in Mexico, commissioned a new computer that was going to run their entire global organization and called it HAL9000, in reference to the film *2001: A Space Odyssey*.[5] The HAL9000 project (Monterrey, 1996–97) is a simple concrete box that has been manipulated to create a powerful identity for the client (figure 2.19). The floor is actually a lighting system. The ceiling provides no light; instead, it is a fabric skin that softly glows with reflected light. As well as being an interesting visual experience, the solution is performance related in that the need for a raised computer floor allowed for accommodating lighting at that level.

2.19
HAL9000, Monterrey (1996–97), architect Grimshaw.

2.20
National Space Centre, Leicester University (1996–2001), architect Grimshaw.

At Leicester University, UK, the practice was asked to design a National Space Centre (Leicester, 1996–2001, figure 2.20) — an enclosure for rockets to excite and inspire the general public, and a base for the university's space research work. The performance objective was fairly simple: creating a large enclosure for the rockets. The challenge was to design a fairly economic space which remained light and airy. We decided to use a material called ETFE foil, which is an air-inflated polymer that can be used to glaze a structure with absolutely minimum weight. This helped to create a unique building, using three-dimensional form to create the architecture and to "consume" the geometries.

Our design for the Messehalle 3 trade fair hall project in Germany (Frankfurt, 1999–2001; figure 2.21) uses geometric solutions in a different way. The roof becomes a piece of origami — a folded plane. Our brief was simple: to design a very large enclosure. The challenge was achieving the 560-foot span. The performance criteria informed the solution. The project team developed a single cord folded-plate roof to cover the span. It is this spanning functionality that gives it form and shape, both internally and externally. Within that large hall many things could happen over the next 100 years — it is impossible to predict how it might be used. A flexible shape was created, but, again, rather than just designing a functional box, by responding to the performance criteria of the brief, the architects gave the Messehalle a sculptural quality that informs the whole nature of the building.

2.21
Exhibition Hall, Frankfurt, Germany (1999–2001), architect Grimshaw.

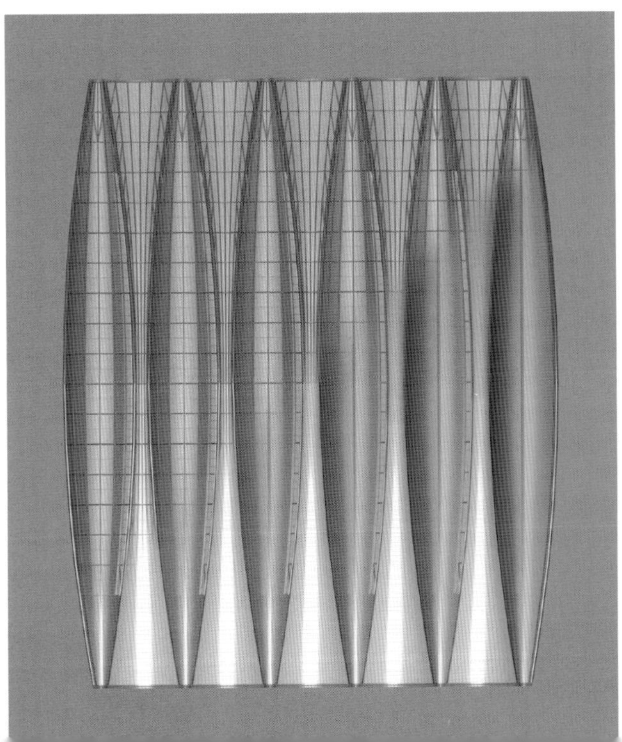

2.22
The Eden Project, Cornwall, England (1996–2001), architect Grimshaw.

2.23
The Eden Project: interior view.

The next three projects show how computers help with three-dimensional geometries, and the relationship between drawing and manufacturing. The first is the Eden Project (1996–2001; figure 2.22) in Cornwall, in south west England. The brief was simply to create large enclosures for plants — a twenty-first century botanical garden. One of the enclosures had to be big enough to allow a rain forest to grow to its full maturity, which meant creating something about 150-feet high by about 300-feet wide in cross section (figure 2.23). That is considerably larger than traditional glass houses.

The site was a quarry, and we thought that half the architecture was in the topography — the quarry's incredible landscape. We captured elements of that landscape to create the enclosures; the architecture should have the same fluidity and a synergy to the topography that it would be nestled in. The initial ideas were similar to those of the Waterloo project — to use a series of nineteenth-century inspired trusses to create a sinuous linear structure that would then be glazed.

Several factors led to a review of this approach, not least the fact that the site was still being quarried and the topography was changing throughout the design process. The radical solution was to create a series of "soap bubbles," sitting lightly in the landscape. Each enclosure was envisaged as a series of spheres made up of hexagonal panels that could adapt to the changing ground plane by adding or subtracting panels as necessary. In practice, this meant creating a computer model of the topography and one of the spheres (figure 2.24), and intersecting them to determine the form of the enclosure.

Cutting through the model gave a changing fluid form as the topography of the ground and form of the enclosure changed. The structure best suited to the resulting geometry was made up of pentagons and hexagons. To minimize the size of the structural sections, we used ETFE foil inflated

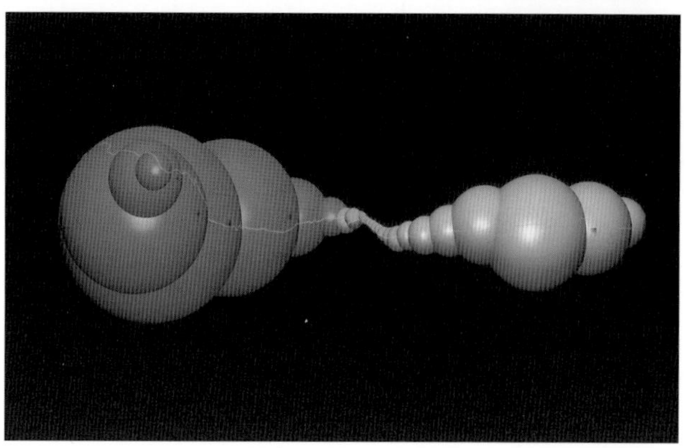

2.24
The Eden Project: the initial spherical geometry.

2.26
The dragonfly's wing.

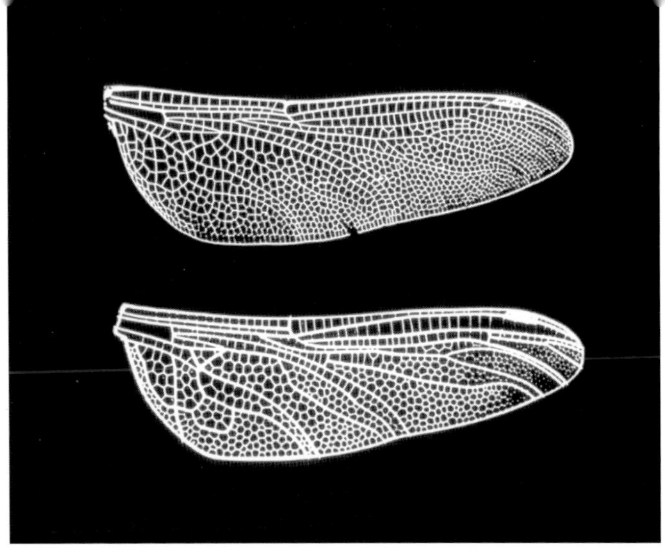

2.25
The Eden Project: interior view.

pillows to glaze the forms — we simply captured air with the structure and the pneumatic skin (figure 2.25).

This structural geometry for the Eden Project was not a new idea. Not only was a model found dating from 1617, it is also prevalent in nature, as it is the most economical way of capturing three-dimensional forms and creating lightweight structures. The complication at Eden was that a series of intersecting spheres was being used and the points of intersection gave rise to broken hexagons.

The Grimshaw project team decided to step back from form and learn more from nature by looking at the way a dragonfly's wing works (figure 2.26). It is made up of panels of skin that consist of hexagons and pentagons (figure 2.27). Where the panels meet, the hexagons break into perpendicular lines, with a seemingly random pattern, but which actually follow the lines of force. The same process was used for the spheres at Eden, allowing the structure to follow the most efficient path rather than imposing some sort of grid form.

In performance terms, ETFE foil lets in more light than glass, including ultraviolet light, so the plants inside can thrive better than they have ever done before in an artificial environment. It was also about a third of the price that a glass and steel enclosure would have been, meaning that the client could afford it — an important consideration.

Most importantly, the enclosure needed to be as light as possible for environmental reasons. If the enclosure was made out of small elements and from a lightweight material, it would require minimal transport to get the system down to that remote part of England, which is poorly served by road and rail. The whole weight of the roof, including the steel and the foil, is no heavier than the mass of the air it encloses. Ultimately, those are just facts and statistics, but overall it is the Eden "biomes" as objects and their functionality that gives them great beauty — the soap bubbles clinging to the rock face.

2.27
The Eden Project: hexagons and pentagons define the geometry of the enclosures.

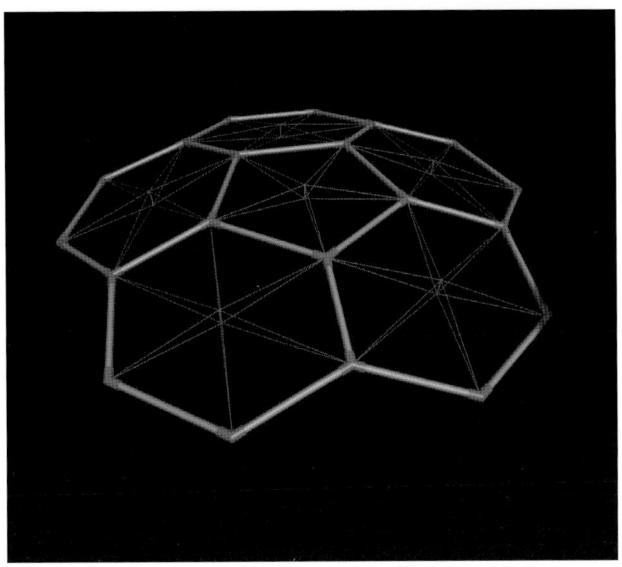

2.28
Southern Cross Railway Station, Melbourne, Australia (2002–), architect Grimshaw.

DIGITALLY-DRIVEN DESIGN AND PRODUCTION

Computers allow architects and designers to follow the laws of nature and to explore things in a much more three-dimensional way rather than being restrained by the means of a set square, a drawing board and a standard manufacturing technique. Computer systems are being developed to describe and define space; manufacturers now own computer-aided design/computer-aided manufacturing (CAD/CAM) facilities to enable rapid prototyping, effectively doing away with notions of mass production. In the Eden Project, for example, that meant that each pillow could be different without any prohibitive cost implications.

Other examples of the impact of CAD systems on architecture include a project that Grimshaw is designing in Australia and a fourth phase at Eden. In Australia, the practice is designing a whole city block that links the Central Business District and the docklands (which are being redeveloped). Within that block, among other buildings, is the new Southern Cross Railway Station (Melbourne, 2002–; figure 2.28). The performance requirements of the roof were a major consideration for the railway terminal. It acts as an umbrella or sunshade, but it needed to be visually interesting as the skyscrapers around look down on it. It also had to extract stale air from the diesel trains. The great fans that provide the obvious solution to this problem, however, are neither sustainable nor aesthetically pleasing. Instead, we looked at the prevailing winds. The wind effectively sculpts and gives shape to the roof in the same way that it creates dunes in sand or moguls in the snow. These dunes force the air to pass over the roof surface, creating negative pressures to lift out and ventilate the space below. So the roof functions effectively but is also visually interesting. It is the performance criteria that give it shape and form — and its sculptural qualities.

On a much smaller scale, Grimshaw is now working on the fourth phase at the Eden Project (Cornwall, 2003–). This phase comprises a series of buildings, including another biome. Again, it required creating something different, pushing the performance criteria a stage further. One of the buildings is an education center for which low-embodied energy construction is being investigated, as 1,000 school children on average visit Eden each week to learn about science and biology; the client naturally wants the building to inspire them. The idea was that the building would tell the story of transpiration, which is the way a tree uses energy. The resulting building is a tree-like shell form that follows a logarithmic geometry in the form of a spiral (figure 2.29).

2.29
Logarithmic, spiraling geometry for the Education Centre, Phase Four, the Eden Project, Cornwall, England (2003–), architect Grimshaw.

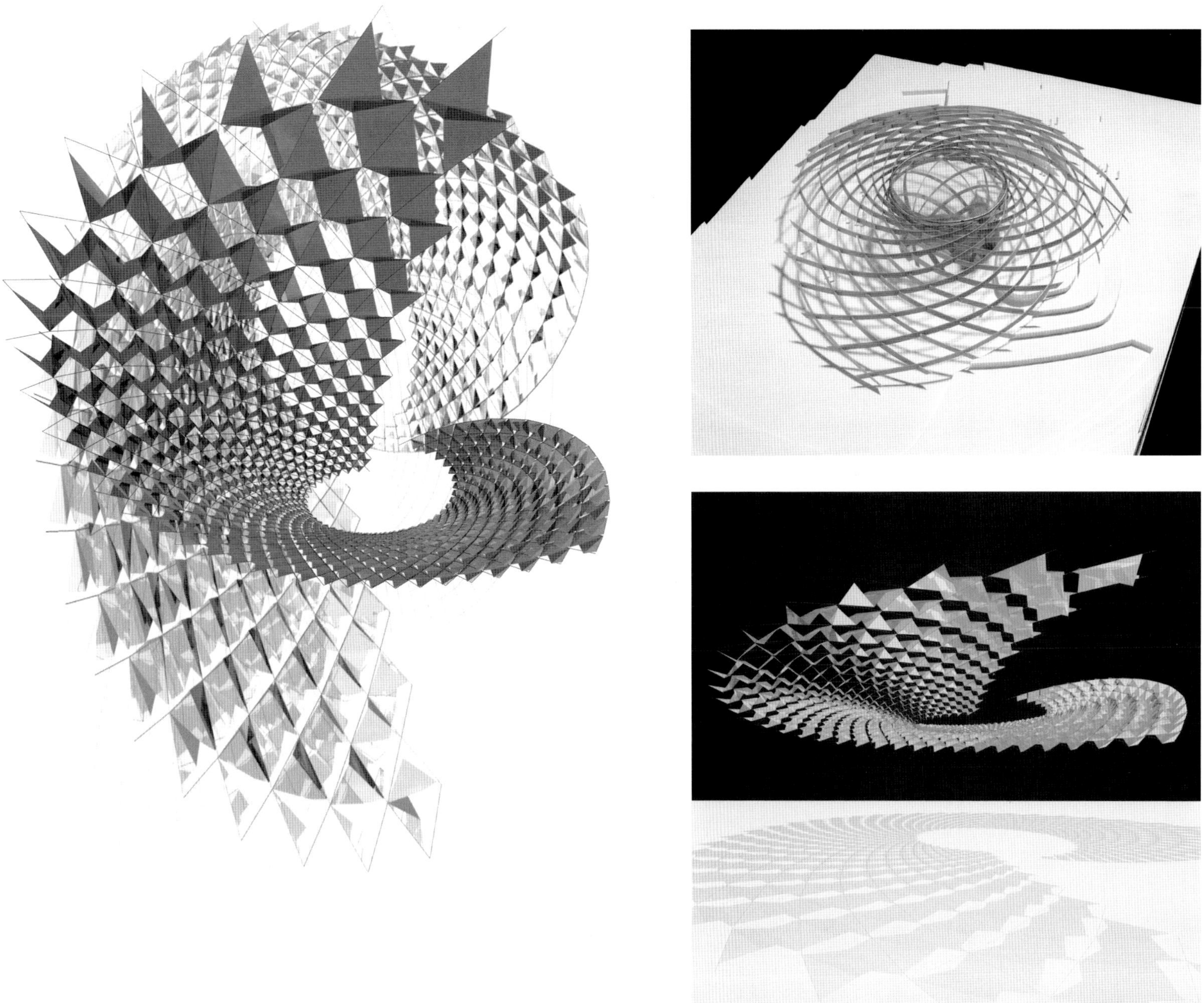

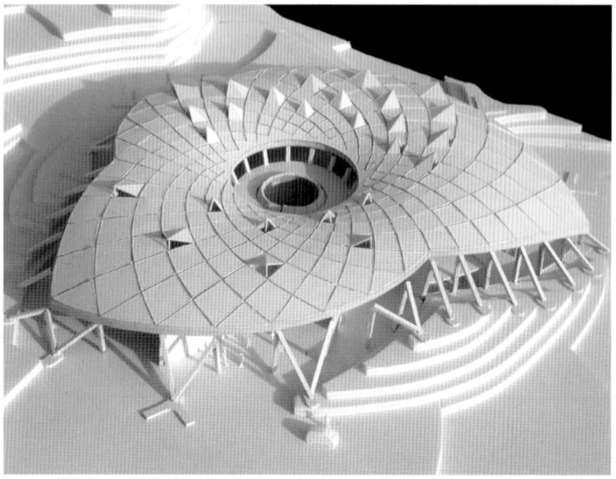

2.30
The Eden Project, Phase Four, Education Centre: a CAD/CAM model.

Although computers play their part, ideas originate in the mind and are worked out with paper and pencil and simple sketch models before they are transferred to CAD. It is only with the computer, however, that you can work out the final geometry, the final connections and the size of elements, etc. The practice had been designing components using rapid prototyping as a way of quickly investigating the shape, performance and function. Now the architects can send a three-dimensional model to a rapid prototyping company and a day later get a model (figure 2.30). It is made overnight, and it offers a great way to investigate form and shape. It can be worked on by hand, adding evidence, thinking about how the skin is going to work, the way it is going to allow the light in, and the way it is going to create a shelter. To that end, Grimshaw has been doing a project with Bentley Systems looking at the way the skin and structure can be described as a series of changeable components and elements. The idea is that everything should be flexible and easily changed.

A new influence on the Grimshaw design process is the practice's Environmentally Viable Architecture (EVA) design guide — a software tool where architects can measure a building's potential impact on the environment. This encourages a performance-led approach to design and architecture, as the architect is given a continual feedback on the impact of their design and materials. The intention is to educate architects and designers to be aware of environmental strategies that can assist in the production of sustainable architecture so that, ultimately, realizing a sustainable solution becomes second nature. Critically, scores are a measure of the process rather than the product.

PERFORMANCE-DRIVEN PROCESS AND PRODUCT

Architects are no longer constrained by the limits of traditional construction techniques; designs can now be fully conceived in three dimensions. More profoundly, architecture can be guided by the same laws that control and shape the world around us — an organic approach to design based on exploring solutions through performance.

The practice is currently working on a project in New York State, a concert hall for the Rensselaer Polytechnic Institute (Troy, 2001–; figure 2.31), developing interesting acoustic performance criteria by using fabrics in a new way. The real questions are: how can Grimshaw design an experimental media center that, by its very nature, is something that changes all the time? How do you meet that challenge architecturally? Who knows what is going to happen in fifty years' time? How can the architects express the potential for change in a functional way, in a "performative" way?

In this example, the answers came from physics. The main concert hall is principally used for performing symphonic work. It is a great acoustic space, and its shape and form is entirely driven by creating the optimum acoustic space. This is the same performative criteria that create a violin's form. A Stradivarius violin is a functional object that produces the most beautiful range of sounds; it is this dedication to performance criteria that results in an object that is also intrinsically beautiful to look at. Through responding to function logically, following the laws of nature and physics, it is possible to achieve something of exceptional quality. The Grimshaw project team wants the concert hall to have permanence through the way it functions, by being shaped and formed through acoustics, like a musical instrument — it uses the law of physics but can produce an object of great beauty in space.

Grimshaw's work demonstrates that good architecture can be process driven. Beauty can come from integrity, if one responds to the functional requirements and designs a truly performance-related space. Following performative criteria does not negate architectural qualities or beauty — when it is done properly, it generates them.

2.31
Concert hall,
Rensselaer Polytechnic
Institute, Troy, USA
(2001–), architect
Grimshaw.

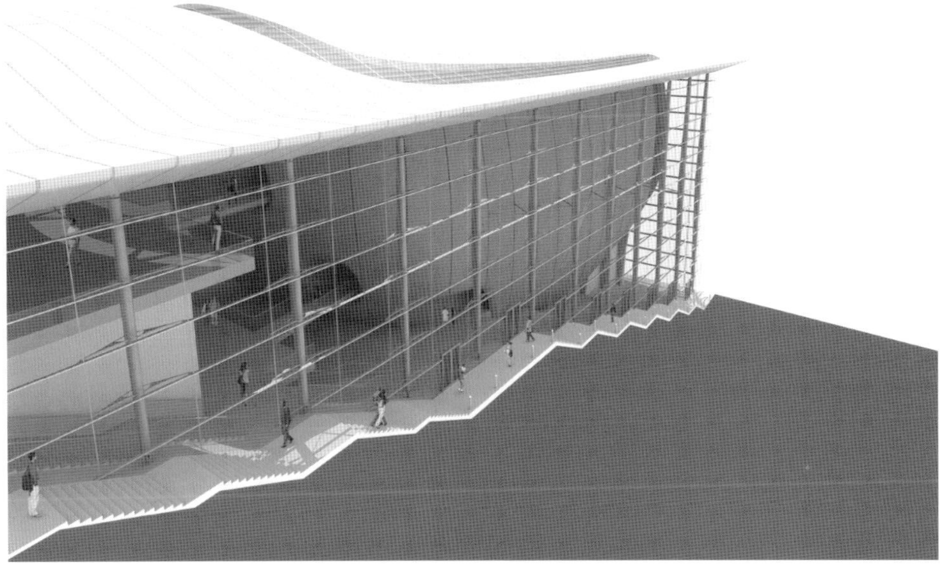

NOTES

1 It is not always easy to predict change. As David Leatherbarrow points out in Chapter 1, the user may alter the use and the building still functions, but maybe in a way the architects never thought about. A practice can try to build into their building designs as much flexibility as possible, but it can never completely anticipate the future.

2 The early buildings were determined by what could be manufactured and the way the practice thought and designed, using the set square and line. The advent of computers has increased the ability to meet targets, to describe, define and explore spaces, and then ultimately to manufacture and make buildings.

3 This project is also a very early example of defining a form fully in three dimensions. It certainly could not have been designed without three-dimensional modeling software, which was used to describe and explore the complex forms in three dimensions.

4 If this was a film then this particular scene ended up on the editor's floor for a series of reasons, but the overall building budget clearly played a part. However, it was a theme that we continued to explore for the new project at the Royal College of Art.

5 Many of the things predicted in *2001: A Space Odyssey* turned out to be true. The film re-evaluates the idea of enclosure. It challenges surface and orientation — what is floor, ceiling, wall, what is volume. It offers a notion of very dynamic, changing environments.

3

SUSTAINABLE DESIGN: AN AMERICAN PERSPECTIVE

MAHADEV RAMAN

3

SUSTAINABLE DESIGN: AN AMERICAN PERSPECTIVE

MAHADEV RAMAN

The American experience with sustainable design differs from experiences in other parts of the world. When we understand the specific conditions that impact on the United States, we can explore whether sustainable design is needed here, where it should be practiced, who can promote such designs and how we can measure their success.

Architects and engineers are exposed to well-known environmental issues, including resource depletion, pollution and global warming. Professionals tout definitions of sustainability but the practical application of these concepts presents major challenges. Sustainability is not just about energy consumption — it is about finding the right balance between environmental, economic and social concerns.

What exactly are we trying to sustain? While much discussion centers on saving the planet, the planet can actually take care of itself. If the human race was eliminated tomorrow, life on earth would continue to evolve and thrive. Therefore, sustainability is not about saving the world — it is about saving ourselves and achieving as much longevity as possible for our civilization.

North America is the highest per capita consumer of fuel in the world. The energy consumed by buildings in the United States represents approximately 40% of this energy consumption. Therefore, the building industry here has an important role to play in shepherding this resource. Sustainability in buildings often means minimizing the consumption of resources (water, energy, materials) but increasingly it also entails maximizing the health, safety and quality of life of their occupants. There are many challenges to practicing sustainable design in our environment.

This chapter explores the phenomenon of sustainable design from an American perspective and identifies the unique opportunities and obstacles designers face when attempting to construct more environmentally responsible buildings in the United States. A number of case studies illustrate the varying degrees of success achieved by the designers at Arup in pursuing a sustainable agenda on projects completed in this country.

CHALLENGES TO PRACTICING SUSTAINABLE DESIGN

In the disposable culture of the United States, there is a sense that everything used can be discarded. Yet, whatever is discarded continues to exist in our one, closed system. Global warming is an example of the detrimental effect that our lifestyles can have on the environment of a closed system. In the United States, where there are about 70 people per square mile, we can continue to create and expand our landfills. That is not so easy in India and the United Kingdom where there are 700 people per square mile. Europe and China have approximately 400 people per square mile. These enormous differences in population density breed varied approaches to the environment. In the United States, we tend to think of our natural resources, including land, water and forests, as unlimited and disposable.

Legislative changes to codes and regulations do not happen as quickly in the United States as in other countries. US citizens value their personal freedom greatly, as do the individual states. Our federal government, for example, converted to the metric system several years ago but since private industry does not need to follow suit, the use of the metric system remains limited.

Code innovation occurs differently in the United States than in Europe. In the United States, practitioners tend to work through the codes and regulatory structure as a means of implementing change. Establishing a higher level of performance in buildings generally needs a consensus for the idea to build momentum so that eventually a code change follows. In Europe, individual practitioners are always pushing the boundaries and this approach seems to be supported by the economics there. However, the downside of this approach is a higher variability in the quality of buildings.

Urban conditions present another challenge. Extensive land resources in the United States have led to urban sprawl, resulting in increased traveling distances, traffic emissions, etc. This condition is less prevalent in more densely-populated areas with public transportation.

Another factor is the country's climatic diversity. There is no single architectural style or engineering solution that is efficient and appropriate for buildings in all climatic regions.

Finally, financial issues shape our approach to sustainable design. Since energy prices for gasoline are approximately 25% of prices in the United Kingdom, the pressure to conserve is not as great. These cost considerations have a similar effect on power conservation. Americans also demand much quicker paybacks on investments than individuals in other cultures.

PROMOTING SUSTAINABLE DESIGN

The stakeholders in a sustainable design argument are many. Owners bear the financial responsibility and have concerns over the long-range impact of a building's design. Occupants have a stake but are rarely involved, so they can be seen as silent stakeholders. Government entities and local authorities need to ensure that their codes are not contravened. Designers and builders, in general, wish to leave a positive legacy. The community itself, both locally and globally, has to live with the consequences of the design for many decades. Finally, there are the future generations of stakeholders, the people that we hope will, in perpetuity, be able to enjoy the quality of environment that we take for granted. Since the majority of these stakeholders are not engaged in the typical design process, the design team has a duty to adopt the positions of others and to view the process from their standpoint.

Certain essential ingredients must be present in any process that aims to achieve a better, more sustainable result. There must be a commitment to the process, particularly from the owner, from the outset. A well-intentioned design team working without the owner's commitment will not succeed.

There must be a non-traditional approach to communication and interaction among the disciplines. A traditional sequential organizational structure will stifle the best results. Communication that allows for feedback and reworking is very important. A brilliant idea may occur spontaneously rather than in a scheduled meeting. A project must have an organizational structure that will allow for the capture of those ideas and the ability to incorporate them into the process.

When a group of designers from different disciplines have collaborated before, they often develop a rapport and a way of playing off each other's ideas, which helps to hone the project. Today's complex, modern building is the creation of many professionals. The best examples of sustainable buildings are integrated in such a way that the engineering is the architecture and the architecture is the engineering. It is the ensemble that creates the desired results.

Also, recognition that various team members (architects, engineers, owners) communicate and process information differently is key for success. The use of a variety of visual aids, animations, and simulations help in this process of communication. Simple diagrams usually prove to be more effective than many pages of text in explaining energy flows and other complex concepts (figure 3.1). Diagrams and visualizations can also be especially effective when communicating across language barriers. When traditional approaches to presenting technical information through tables and numbers are brought to life through animation, it becomes easier to synthesize large amounts of data, not only for the team members and code officials but for the presenter as well.

The final essential ingredient is a commitment to properly follow through. Typically, not enough is done to ensure that the design intent is being met in all of its depth. If there is not a substantial commitment to follow through in the process, then the best design intentions will not be realized in practice.

MEASURING SUCCESS

Limited budgets and fee structures are practical constraints that place a limit on the sustainable goals that can be achieved. Once stakeholders are identified, educated and committed, targets that seriously consider budgets and fee structures must be set. These targets can be very simple, such as energy targets, and we can use grading systems such as LEED (Leadership in Energy and Environmental Design) to measure progress in meeting targets.

LEED is being adopted as a means of measuring a building's sustainability. Does a LEED Platinum rating indicate that a building is truly sustainable? Not necessarily. Rather, it indicates that the design team has achieved a certain amount of points on the LEED rating scheme. While LEED is an excellent tool and provides some focus to a design team, the

3.1a
Sunwall Sketch, winning entry for the Department of Energy competition located at their building in Washington, D.C. (2000), architect Solomon Cordwell Buenz and Associates.

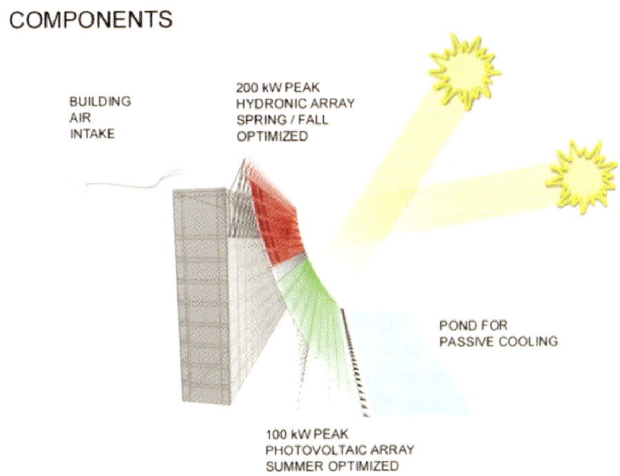

3.1b
Rendering of the design for the American University in Cairo, Egypt (1999, not built), architect Sasaki Associates.

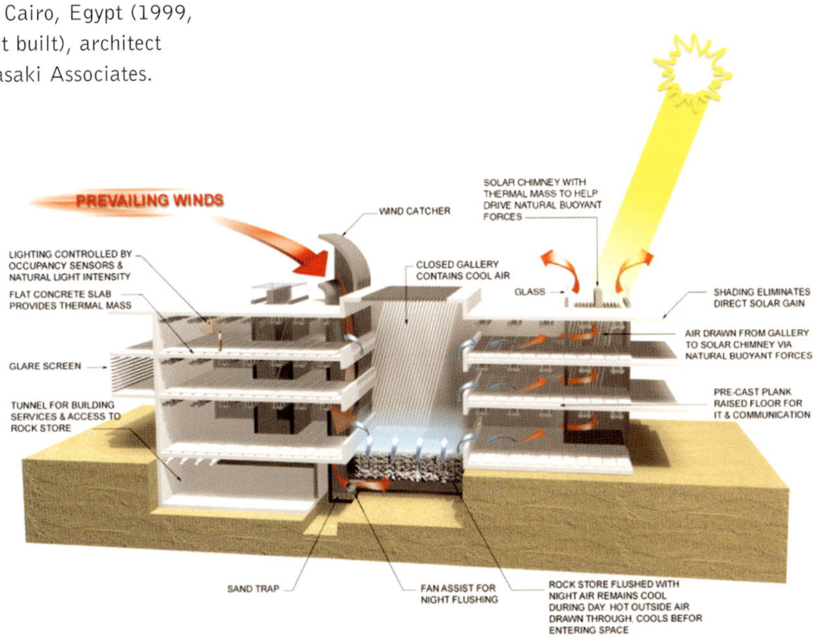

system needs improvement and broadening. Setting targets and securing points will not take the place of a good design, but a good design team can take targets and turn them into amazing results.

COST CONSIDERATIONS

While first costs are often the foremost consideration in building projects, operating and refurbishment costs need to be considered as well. How often do designers think about the cost of removing and replacing a piece of equipment in 25 years' time? A smart design approach, which tries to address the varying lifespans of the different building elements and systems, can significantly facilitate future refurbishment.

Fees in the United States, particularly for engineers, are lower than elsewhere. But salaries for those same professionals are higher. Consequently, the design process in the United States may use half to one-third of the man-hours devoted to the design process in the United Kingdom, for example. That is a staggering difference, which seriously limits the amount of time that can be spent to refine designs and move them to a higher level.

The linear process is a natural response to the pressure to complete the design in fewer man-hours. In addition, the fact that schedules are tighter in the United States also leads to a more regimented process. To produce smart, sustainable designs within these constraints, designers need to be disciplined and focus efforts more at the early schematic design stage, and to test options while there are fewer constraints on the process. Once a smart schematic design is in place, the design needs to follow the more linear process because it is not economically feasible to work otherwise. Innovation must be quick and early.

A sustainable design process, that is to be done properly, requires higher levels of professional fees than a conventional process. We can estimate that the schematic design may need 40% more effort and the design development perhaps 20% more. It may not require more effort to produce innovative construction documents but more follow-through is imperative. Across a project, we can estimate that 15–20% more fee is needed to deliver an innovative sustainable design.

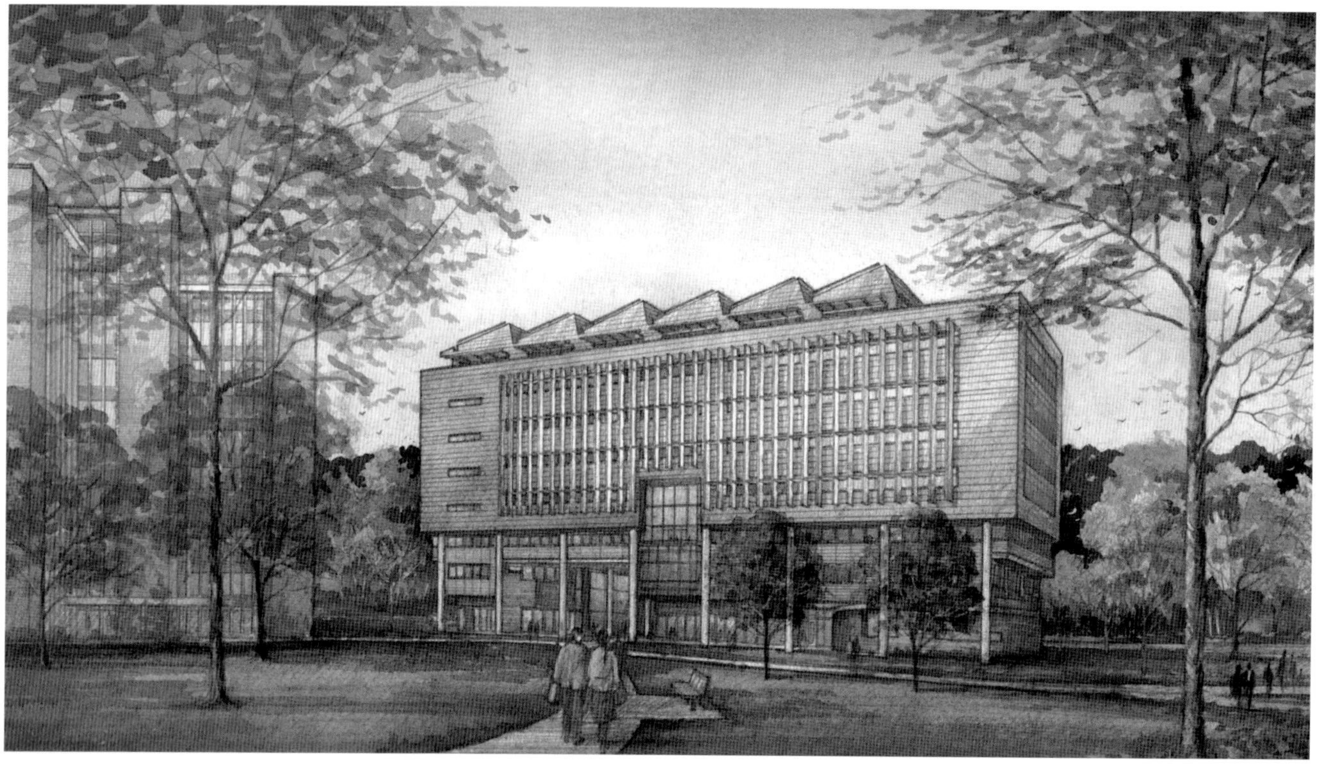

3.2
Rendering of the University of Texas Nursing School (currently under construction), architect BNIM Architects.

CASE STUDIES

Owner commitment

A client from the University of Texas proved his commitment to environmental issues by replacing a project's first architect when he felt that not enough was being done to incorporate environmental features. While the resulting building is not outwardly remarkable to look at (figure 3.2), it offers a series of interesting features in terms of water recycling, solar harvesting, appropriate orientations to minimize solar heat gain, and superb use of daylight throughout. This building is going to achieve a LEED Platinum rating and the client is now pleased.

The Gap is committed to presenting an environmentally conscious and friendly image, and wanted to embody this philosophy in their headquarters building in San Bruno, California. The site selected by the Gap was inspiring because it sits in a climatic zone where buildings can be naturally ventilated with relative ease. In the resulting building, air-conditioning loads are low and the design incorporates a variety of progressive features. Among the features is a grass roof that is designed to provide a certain amount of photosynthetic replacement for what has been taken away from the site (figure 3.3). In the workspaces, no person is more than 40 feet from a view and a source of daylight. Spaces like the atrium enjoy pragmatic approaches to daylight. The orientation of windows is vertical because vertical windows are easier to seal. A curved sheetrock surface efficiently reflects light down and into the office space (figure 3.4). If the LEED system was in place when the Gap headquarters was constructed, the building would likely have achieved at least a Gold rating. The Gap now has very high employee retention rates and this building has probably already paid back any additional costs that may have been incurred by making sustainable design choices.

3.3
GAP Headquarters, San Bruno, California (1996–1998), architects William McDonough + Partners and Gensler.

3.4
Interior of the Gap headquarters.

Non-traditional approach to communication and interaction among the disciplines

The team organizational structure for a residence hall at the Massachusetts Institute of Technology (MIT) in Cambridge, USA (1999–2002, architects Steven Holl Architects; figure 3.5) fostered a multidisciplinary, non-linear process that required feedback. The owner community had four distinct constituents, each with its own driving mission. The MIT President desired a landmark building in terms of architecture and sustainability; one goal was to develop the building without traditional air conditioning. MIT's project management team had a clear mission to deliver the project on budget and on time. The facilities group needed assurance that residents would be comfortable. Members of MIT's Building Technology Faculty did not want to be associated with a naturally ventilated project because they did not believe the goal could be achieved. The design team included two architectural firms, two structural firms, two Arup offices and a host of other consultants.

3.5
Simmons Hall at MIT, Cambridge, USA (1999–2002), architect Steven Holl Architects.

3.6
Simmons Hall: Atrium Student Lounge computational fluid dynamics (CFD) study.

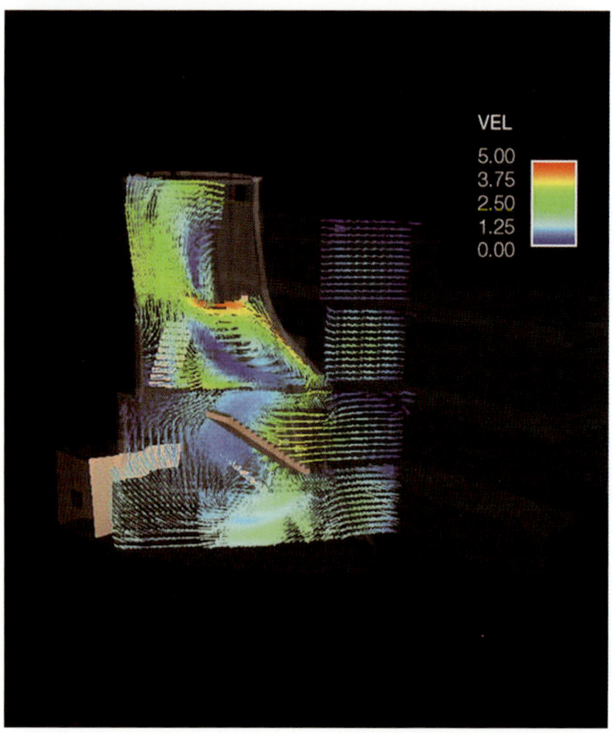

While satisfying the differing interests of this group was a considerable challenge, the end project is a tremendous success. In the plan that emerged, architecture, structure and building systems developed as one integrated whole. All building elements were designed to minimize the cooling load. The façade consists of 3,000 windows (figure 3.5), each set deep within an 18-inch perforation that allows for low winter sunlight to enter while shading the rooms from the high summer sun. The thermal mass of the façade soaks up the solar heat, thereby minimizing the thermal plume that could enter the rooms via open windows. This design enabled Arup to cool the building by way of a mixed-mode system that provides a very small quantity of dehumidified cooled air whenever the outside temperature rises above 80°F (figure 3.6). There have been no complaints about occupants being too warm.

A project at New York's Penn Station (1998–2003; with Skidmore Owings & Merrill) illustrates the success of visual aids. Designers studied the effect that light coming in through an expansive skylight would have on electronic display boards being proposed for the station. If the

3.7
Lighting simulation, Penn Station, New York, USA (1998–2003), architect Skidmore Owings and Merrill.

3.8
Lighting simulation designed to investigate the visual clarity of electronic displays for proposed Penn Station passageway.

brightness of the light from the skylight could be controlled without destroying the architecture of the space, then designers could specify less bright and, therefore, less expensive display boards. Arup not only used a rendering package but also a photometrically accurate simulation to measure daylight, reflections and other criteria (figures 3.7 and 3.8). Designers that are capable of such an advanced simulation can conduct these types of studies at the design stage, make decisions early, and move forward with a greater level of confidence — an immensely valuable approach. Buildings can be constructed, examined and tested virtually before a builder sets foot on the site.

At Norman Foster's Greater London Assembly Building (1999–2001), a particular communication problem was solved in an innovative way. The initial scheme for the assembly hall was a very smooth form. When the architect envisioned the acoustics, he imagined that the sound of a person's voice would bounce up and escape out at the top of the shape. But acousticians explained that the assembly hall would be excessively reverberant and that, without some absorption measures, the acoustics would be terrible. Arup developed a process to visualize the sound being reflected and absorbed on different surfaces, and, after several iterations, a solution emerged that was both architecturally and acoustically acceptable (figure 3.9).

3.9
Acoustic progression, Greater London Assembly, London, UK (1999–2001), architect Foster and Partners.

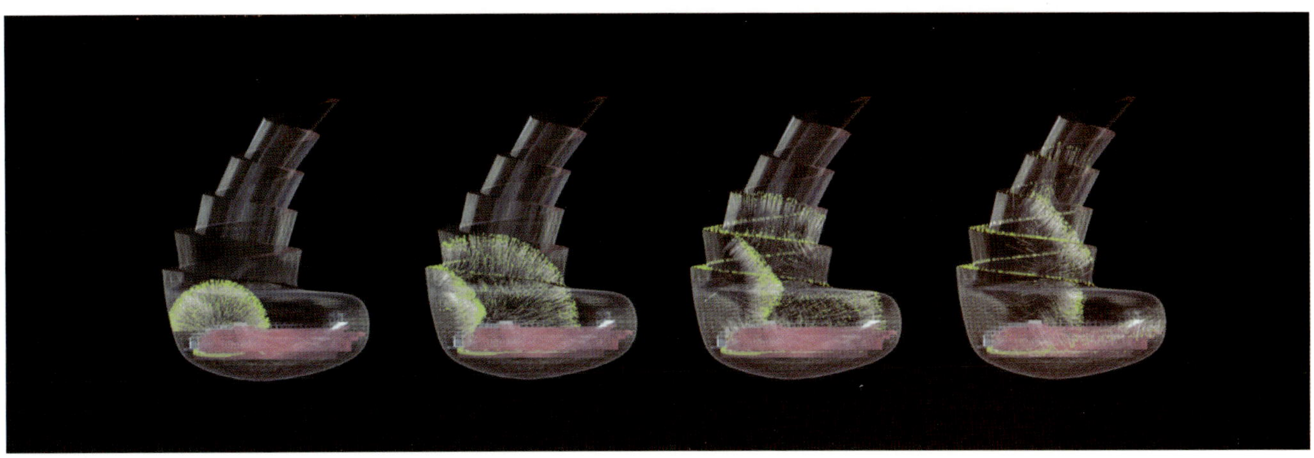

3.11
Terminal 4 at JFK Airport, New York, USA (1996–2001), architect Skidmore Owings and Merrill.

3.12
JFK Terminal 4: linear skylights provide daylighting.

3.10
Simulex evacuation pattern.

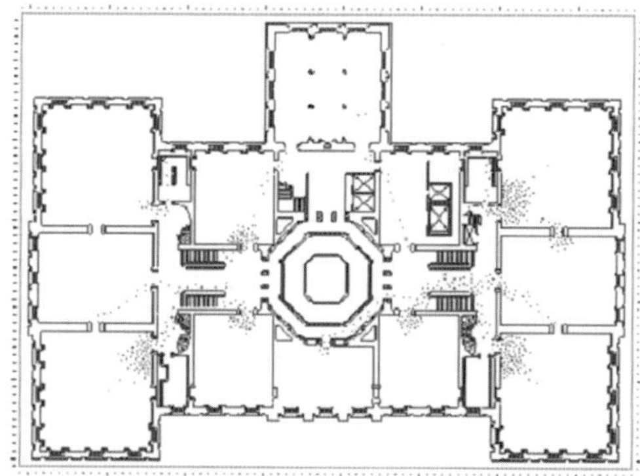

Simulations can be applied to many design conundrums. One particularly challenging issue is designing escape routes in complex environments, such as, for example, large transportation buildings with multiple levels. In these unconventional buildings, it is often difficult to include multiple stairways to meet the letter of the code. However, the openness of these buildings often means that occupants can find safe escape routes with ease during an emergency. One powerful tool to demonstrate building safety is a computer program called *Simulex* which enables designers to run an emergency evacuation simulation that considers demographic data of building occupants, including age and agility. The program produces very specific real-time animations that demonstrate the adequacy of escape provisions and accurately estimate the amount of time necessary for all occupants to escape from the building (figure 3.10).

Commitment to follow through

Site supervision, commissioning, handover to the facilities group and the management of a facility are the keys to achieving positive results. Arup's client for Terminal 4 at JFK Airport in New York (1996–2001, with Skidmore Owings and Merrill; figures 3.11 and 3.12) fully understood these keys; the same client was responsible for funding the project, building it and operating it for an extended number of years. The client paid the design team to maintain approximately thirty on-site personnel throughout the construction to ensure that the design intent (including an advanced integrated IT system) was fully realized. The team was able to immediately troubleshoot any obstacles to achieving performance goals during construction. It then oversaw the painstakingly detailed commissioning of all systems. In Terminal 4's first year of operation, the client is reaping the benefits of its investment. While the building is approximately twice the volume of the terminal it replaced, it uses only about 60% of the energy.

3.13
Lerner Student Center, Columbia University, New York, USA (1994–99), architect Bernard Tschumi.

3.14
Close-up of exposed steel pedestrian walkways at Lerner Student Center.

Measuring success and cost considerations

A major issue at Columbia University's Lerner Student Center in New York (1994–99, with Bernard Tschumi; figures 3.13 and 3.14) was exposed steelwork. The steelwork was critical to the architecture but the code required it to be fireproofed. Arup conducted a fire engineering study that illustrated there was no credible threat that fire loads could compromise the structure during the time it would take occupants to escape. A strong case was presented to the building department and, after careful study and a period of technical arguments, the building department accepted Arup's position. The study saved approximately $750,000 in intumescent paint costs; intumescent paint is notoriously detrimental to the environment.

3.15
Federal Courthouse, Phoenix, USA (1995–98), architect Richard Meier and Partners.

3.16
CFD thermal study of the Phoenix Federal Courthouse.

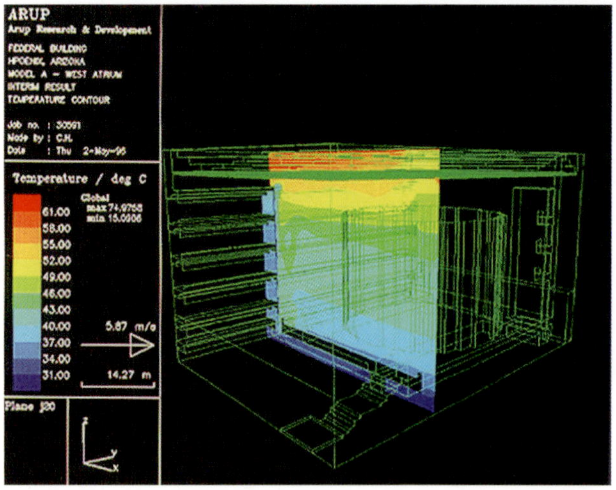

Richard Meier's design for the Phoenix Federal Courthouse (1995–98; figure 3.15) involved the construction of a glass box in the Arizona desert. Using computational fluid dynamics (CFD) and other thermal simulation techniques (figure 3.16), Arup devised an evaporative cooling system that conditions just the base of the atrium where the people are located. The solution cut approximately $5 million off the cost of conditioning the space and reduced the operating costs by approximately $50,000 annually. The design, however, is not necessarily a paragon for sustainability. Arup took a space that would be intrinsically hot and would require enormous amounts of energy to air condition, and instead introduced water as a means of creating an acceptable comfort level for the occupants. Normally, water would be considered a scarce and precious resource in a desert climate. However, with long distance pipelines and massive pumping systems, our society is able to supply millions of gallons of water to Phoenix at a very

3.17
United Nations Headquarters, New York (1947), with the Chrysler Building in the background.

low cost. In this case, it is difficult to compare the relative environmental impact of reduced energy consumption with that of increased water consumption in a desert environment. In the LEED system, a building earns points for saving energy and water, but the LEED system does not address environmental comparability and regional climatic issues. In the absence of concrete answers and guidelines, designers need to ponder these issues and use their best judgment.

Historical approaches to energy and technology

The Chrysler Building (1930) was designed as an extension of the conventional punched window masonry façade building with no concern for air conditioning. The Empire State Building (1931) has a similar façade but introduced early air-conditioning systems. A new approach is evident in the United Nations Headquarters (1947; figure 3.17), with its clear single-glazed curtain wall. Here architects were "liberated" in their approach to design by the ability to install air conditioning. As part of a capital master plan, Arup conducted a survey of the property and revealed the extreme care taken by both designers and builders on the project. The United Nations Headquarters is three-eighths of an inch out of vertical over the entire 36 stories and the stone's condition is amazing. This is attributable in part to the architects' approach to the stone. The architects selected a stone that has pockets on its surface which hold moisture. To eliminate concern over freeze-thaw damage, the façade is designed with a large heating coil on the back of the wall — an approach that illustrates the disregard for resource conservation at the time!

Part of Arup's master plan brief was to study how to improve the façade. Curtain wall panels were removed in several places where damage and corrosion were most likely to occur. Surprisingly, the building had been built so well that most components were like new. Arup reported that, on the basis of the condition of the curtain wall alone, there was no justification for changing it. Arup then looked at the energy issues and found that the cost of replacing the clear single-glazed panes with high performance double-glazing would take over 25 years to achieve a payback in terms of energy savings. This clearly shows that, while high performance facades are environmentally sound, they are actually not cost effective to use in the United States, based on current energy prices.

CONCLUSION

The future is bright for sustainable design in the United States. It will be even brighter when a clear connection can be drawn between environmentally sound buildings and higher occupant productivity.

People intuitively understand that being in a better environment makes one feel better and more productive, but it can be very hard to put numbers to it. While the connection is difficult to measure, there are some recorded cases. Randolph Croxton designed a new day-lit factory for VeriFone adjacent to a conventional factory. The two buildings have the same kind of area and house similar activities. The old building is a dumb opaque shed with industrial lighting while the new building has a lot of daylight. The productivity in the day-lit space is much higher and people fall ill less. Apparently there is a long waiting list of people to move from the old to the new building. In this case, the productivity gain is clearly demonstrated, and the economic benefit to the business can be analyzed and documented.

As concrete evidence mounts and as people begin to understand the true value of environmentally sensitive design, opportunities to fund, design and construct well-integrated buildings will increase. It will become practical for owners and stakeholders to invest the time and financial resources required, both for early-stage planning and, critically, for on-site follow-through.

The current challenge, then, is to demonstrate the direct relationship between the environmental quality of a building and business profits. A 1% increase in staff productivity can easily justify a 5–10% increase in the construction budget. Owners will come to understand that in addition to the moral case for sustainable design, encouraging environmentally progressive design can also be a smart business move.

4

BIOTECHNIQUES: REMARKS ON THE INTENSITY OF CONDITIONING

WILLIAM BRAHAM

4

BIOTECHNIQUES: REMARKS ON THE INTENSITY OF CONDITIONING

WILLIAM BRAHAM

PERFORMANCE DESIGN (AGAIN)

In 1967 *Progressive Architecture* magazine published a special issue on "performance design," explaining it as a set of practices that had emerged from general systems theory, operations research and cybernetics thirty years earlier, at the end of the World War II.[1] The editors described its practitioners as "systems analysts, systems engineers, operations researchers" and argued that it was a more "scientific method of analyzing functional requirements," which involved "psychological and aesthetic needs" as well as physical measures of performance. The interest in performance clearly draws on the long history of determinism and functionalism in architecture, understood in large part through the mechanical and organic analogies of the late nineteenth and early twentieth centuries. It is perhaps fitting at the outset to recall that Le Corbusier's famous description of a house as "machine for living" was his adaptation of the phrase that he and Ozenfant had earlier used to describe painting, a *machine a émouvoir* — a machine for moving emotions. All the objectivity of functional methods depends on the assessment of subjective needs, of quantified and temporarily stabilized desires.

To enter the discussion of performance design (again), this chapter examines the environmental performance of contemporary buildings. In the last half-century, buildings have become bigger in a new and bulked-up sense; they enclose ever larger volumes, which have been engineered for ever greater comfort and productivity. This bulking-up of modern construction has been made possible by its systems of conditioning — air conditioning, artificial illumination, plumbing, electric power, telecommunications, and now networked information flow — which allow them to assume radically new scales and configurations.

To describe the mechanisms underlying the intensification of conditioning, I have adopted Frederick Kiesler's provocative term "biotechniques," with all its implications of equivalence between biology and technology. In the current context, biotechniques might best be described as the biological analysis of technological systems. They represent the collapse of the mechanical and organic analogies in architecture within the powerful concept of complex system dynamics. The intensification of conditioning operates equally on buildings and their inhabitants, literally conditioning them to want and then "need" the new services, and steadily escalating the levels of comfort and convenience they expect. That process has its thresholds of intensity, beyond which results can be both unexpected and difficult to reverse.

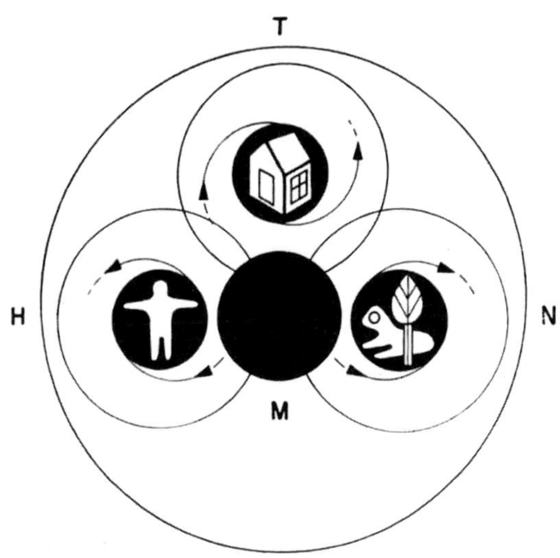

4.1
"Man = Heredity + Environment. This diagram expresses the continual interaction of both the total environment on man and the continual interaction of its constituent parts on one another." From Kiesler, "On Correalism and Biotechnique."[6]

BIOTECHNIQUES

The term "biotechniques" was coined by the architect Frederick Kiesler in 1939 to indicate the equivalence between biology and technology.[2] He was affiliated with Buckminster Fuller's Structural Studies Associations at the time and he used the term to distinguish his thinking from the more direct imitation of biological forms or processes, which today we call biomimicry, and was being called biotechnics by Patrick Geddes, Louis Mumford and Karel Honzik in Kiesler's time.[3] As he observed in an acerbic footnote, "[the Crystal Palace] was built by Paxton in 1851 in imitation of the African water lily's foliate, with its longitudinal and transverse girders. This was an essentially romantic attempt to fashion a man-built structure by literal application of nature's design principles."[4] Instead, Kiesler based his term on a concept he called "correalism", by which he meant "the dynamics of continual interaction between man and his natural and technological environments."[5] I do not mean to claim Kiesler as the originator of these ideas — they were being explored in many fields — but he saw earlier than others how radical their implications were for architecture. Those implications derived from three basic propositions: first, that technology was based on steadily evolving human needs; second, that despite their origin in human needs, technological systems develop according to their own "laws of heredity;" and third, that the final criteria of technological design is not technical performance, but human health (figure 4.1).

In my adapted usage, biotechniques are any method by which buildings are examined as participants in dynamic, "living" systems, whether of the biosphere or of financial, technical or social systems. They may or may not produce results that look biological, and they were initially deployed metaphorically to explain or understand how buildings or artifacts changed or adapted through time. Such biological analogies became more substantial with the introduction of devices and systems that literally flowed or operated — plumbing, electricity, heating, ventilation and lighting — especially with the introduction of feedback techniques, like thermostats, CO_2 sensors and daylight monitors, that enabled building systems to adapt and respond independently. As these elements were fixed in products, codes, standards and procedures, the building of flows and its feedback devices became the legal norm, while new techniques emerged to understand and regulate the dynamic aspects of design.

Such biotechniques became ever more important in the decades after World War II, as cybernetics and general systems theory were applied to organisms and artificial systems alike, rapidly collapsing the difference between mechanical and organic analogies, and making both increasingly operative. This is a critical point. At the moment that living organisms (or ecologies) are understood as kinds of feedback systems, then the difference between mechanical and organic systems virtually disappears. And almost from the beginning of systems research, natural and artificial systems were analyzed together.[7] The career of Jay Forrester, who developed the World III model used in *The Limits to Growth*,[8] exemplifies this process. After early work on air defense systems, he focused his efforts on *Industrial Dynamics*,[9] evaluating the dynamic problems inherent in industrial production, sales and advertising, such as seasonal cycles, countercyclical policies, price stability, sensitivity and unexpected responses to all manner of events, actions and decisions. Through a chance meeting with an ex-mayor of Boston, he applied the same techniques to *Urban Dynamics*,[10] and then after a conversation with the Club of Rome applied them to *World Dynamics*,[11] exploring the interaction between population, industrialization and pollution. This kind of world and climate modeling was central to the developing awareness of global environmental effects, making the construction and authorization of such models vital (figures 4.2 and 4.3).

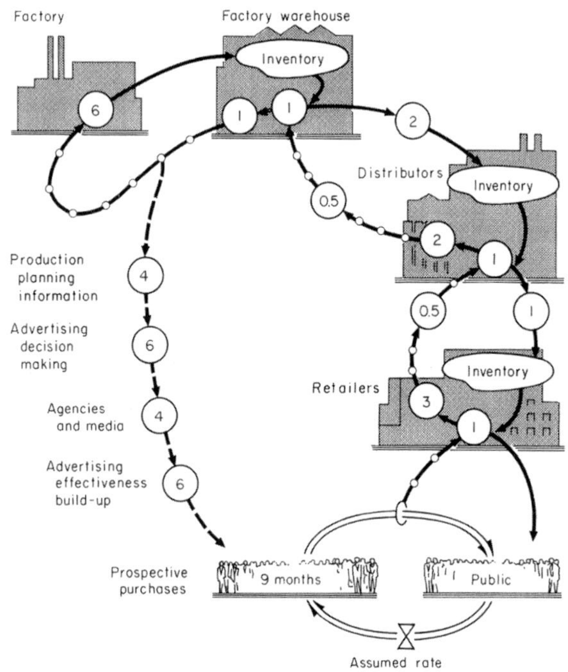

4.2
Simulation model of "production-distribution system," from Forrester, *Industrial Dynamics*.[12]

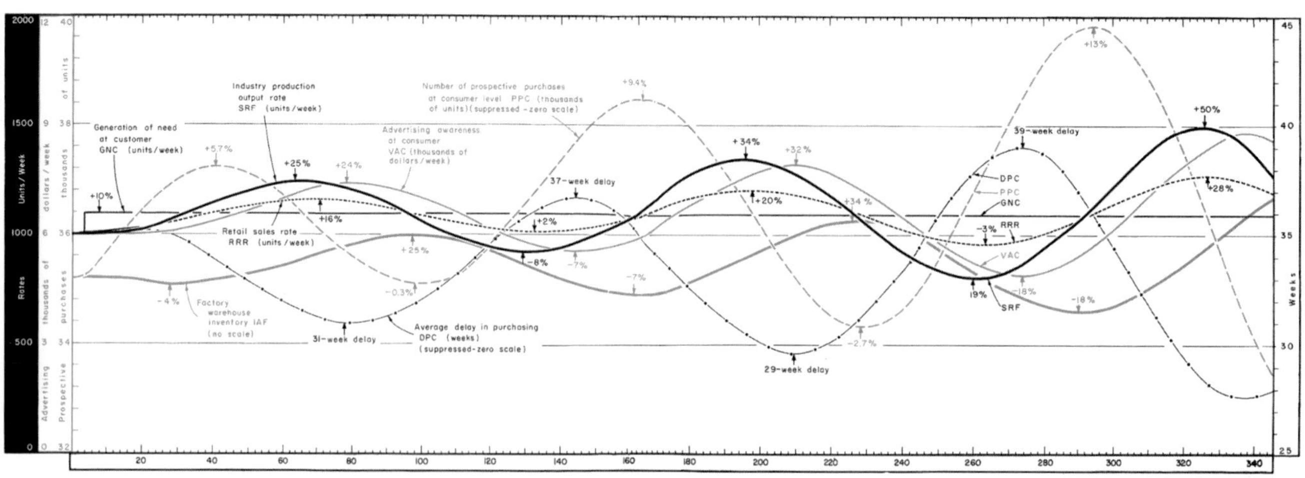

4.3
Amplified oscillatory response of sales and factory inventory to the introduction of a feedback mechanism in advertising, from Forrester, *Industrial Dynamics*.[13]

There have been many criticisms of these simple models, mostly that Forrester's results exceeded the precision of any data that were available. In defense, he argued that the "interaction between system components can be more important than the components themselves" and that the "computer model embodies a theory of system structure."[14] In *World Dynamics*, his primary interest was global population and the early models captured were the dynamic, non-linear effects of multiple feedback conditions; the effect of pollution, food production and resource shortages on population and then of population on food, pollution and resources. But like the contemporary simulations of artificial life, what these simulations lacked were any of the surprising and innovative developments that seem to characterize actual events, or even the internal "laws of heredity" of technological systems. They could not simulate the unpredictable effects that occur at certain intensities of population, such as occurred in the political transition from city-state to national political organization or in the technological transition from wood to coal, oil and gas.

The power of such models lies in their demonstration of effects that are complex, non-intuitive or disproportionate to our actions. For example, many kinds of traffic jams occur once a certain number of people decide to drive, once a certain threshold volume of cars is on the highway. The creeping or stop-and-start traffic that results is not caused by any one person's speed or decision to drive, but occurs like the change of phase as a freely flowing liquid congeals into a solid at a certain temperature (and pressure). One of the greatest challenges for environmentalists is to demonstrate the connection between seemingly minor individual actions — driving to the supermarket, turning on an air conditioner — and these kinds of threshold effects. And if the model is more important than the data, the question for any dynamic simulation is what flows and connections to model? As Forrester's early work suggested, the critical sources of environmental problems are ultimately social, cultural and political, deriving from ideas about health, wealth and pleasure.

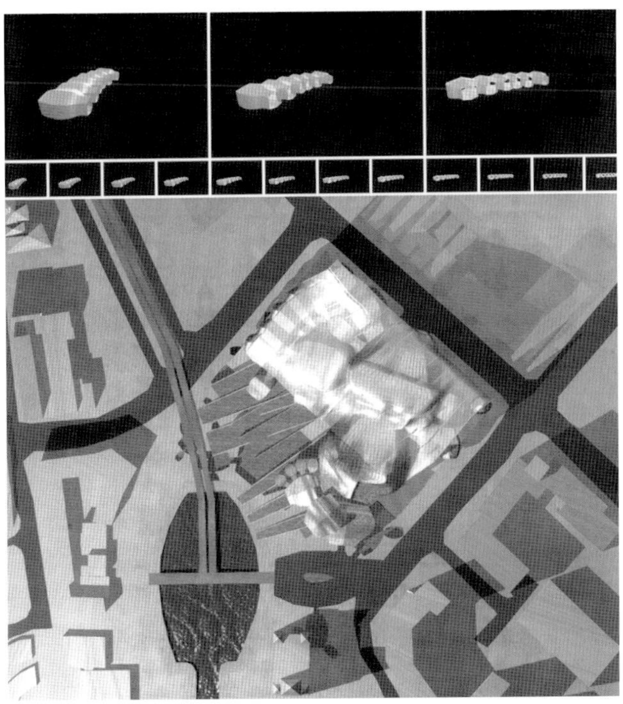

4.4 Dynamic, morphogenetic design model, Greg Lynn, Cardiff Bay Competition, 1994, from *Animate Form*.[16]

BIOTECHNIQUES: MORPHOGENETIC PRACTICES

For over a century, architects have sought "organic" techniques for generating building form, deriving them from structural diagrams, from charts of function, and now from flows of data made manifest with digital animation software. The interest in these new techniques is not difficult to assess. In a 1996 article entitled "Blobs (or Why Tectonics is Square and Topology is Groovy)," Greg Lynn argued that "the mobile, multiple, and mutable body, while not a new concept, presents a paradigm of perpetual novelty that is generative rather than reductive."[15] The novel morphogenetic properties of the new models are made possible by the development and animation of "'isomorphic polysurfaces' or what in the special-effects and animation industry is referred to as 'meta-clay,' 'meta-ball' or 'blob' models." Lynn explains that "in blob modeling, objects are defined by monad-like primitives with internal forces of attraction and mass. Unlike conventional geometric primitives such as a sphere, which has its own autonomous organization, a meta-ball is defined in relation to other objects. Its center, surface area, mass, and organization are determined by other fields of influence." Those "fields of influence" can be used to simulate anything from the motion of the sun to the movement of people to changing brand identities, anything whose influence can be assigned a value (figure 4. 4).

Critics like Michael Speaks have noted the apparent contradiction between the responsive dynamism of these animate models and the inherently static nature of buildings.[17] Speaks used his critique of novel and autonomous form to ask for a more flexible form of practice, in effect, opening design processes like that

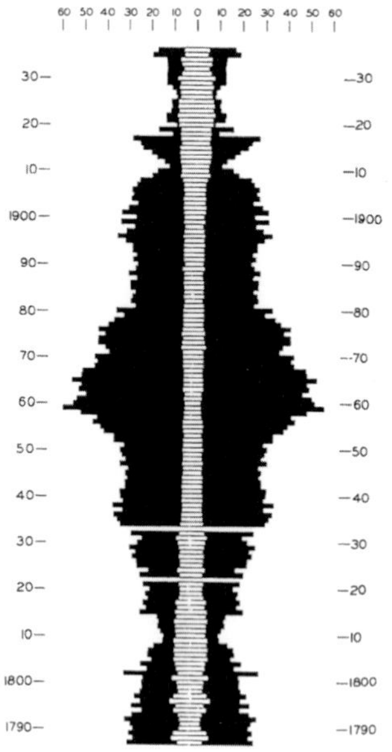

Fig. 6. Skirt width and waist width, dimensions 5 and 6, showing generally inverse relation, 1787-1936.

4.5
Dynamic variations in women's dress dimensions from 1787–1936, by Kroeber and Richardson, "Three Centuries of Women's Dress Fashions."[19]

described by Lynn to the fluid demands of the market. One could certainly point to many forms of architectural practice that have adapted quite aggressively to market forces, from corporate design-build to the absorption of professional designers into large companies. But the critical aspect of these morphogenetic practices lies in the use of explicit techniques to describe the flows, forces or elements influencing the production of buildings. For the most part those designers have wisely avoided the fully deterministic conclusion of their techniques, using them as generative components in otherwise conventional design relationships. But a few historical examples suggest just how challenging a fully dynamic account of design might be.

In 1940, the distinguished anthropologist A.L. Kroeber and a recent student of his published an article on the "quantitative analysis" of "women dress fashions."[18] They charted the skirt and waist dimensions of women's fashions over three centuries, producing some fascinating diagrams that showed the tendencies, trajectories and limits within those basic clothing forms. Certain limits are physical — dresses can only get so wide or narrow — while the basic trajectories appear to have their own momentum, like Kiesler's technological "laws of heredity." Kroeber and Richardson were careful not to speculate beyond their data, but, as with architecture, it is quite common to imagine that the changes in women's dresses would correspond to events outside the fashion system, to wars, economic events or the weather, expressing a certain spirit of their time. What that classic approach neglects is the degree to which those trajectories and their momentums are constrained by the dynamics of the fashion business itself — its techniques of production, marketing and sales — and ultimately by the collective changes of taste. And even conceived of as one of Forrester's dynamic situations, such an account cannot predict when new possibilities emerge, when women begin to wear pants, for example, or when some women wear thin skirts, while others wear wide ones (figure 4.5).

To carry the analogy to its conclusion, new clothing possibilities emerge at different kinds of thresholds, when the pace of fashion accelerates beyond a certain point or when too many women (and men) are participating in the fashion system. In his 1960 book on the planning of shopping centers, Victor Gruen sought to illustrate the synergistic conditions that enable a new shopping center to emerge, to understand the necessary "chain reaction between investment, income and financing."[20] While the analogy was drawn from physics, the dynamics implied are thoroughly ecological. The emergence of a successful shopping center is explained as a delicate interaction between factors like the "financing climate, economic climate, business potential, management skill, and general cost level." He used the analogy and his decades of experience to describe target values for those factors but, of course, this model would only describe the emergence of the form with which he was familiar, not of something different, like big-box retail (figure 4.6).

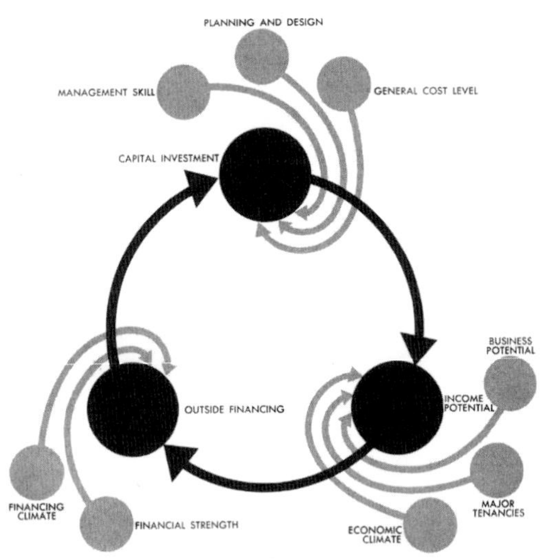

Secondary factors in the chain reaction.

4.6
Primary and secondary factors in the "chain reaction" of shopping mall development, from Gruen and Smith, *Shopping Towns USA,* 1960.[21]

BIOTECHNIQUES: BUILDING PRODUCT INFORMATION

For most buildings the critical flows are neither energy nor resources, but money and product information. That situation is exemplified by the ever expanding *Sweets Catalog* and the whole messy system of selling building materials, products and processes. Sweets originated in the 1890s as a service of F.W. Dodge Construction.[22] The first full catalog appeared in 1906, with an introduction by Thomas Nolan in which he "very gladly consented to commend the idea [of] a really *scientific standard catalogue and index of building materials and construction."* He explained that he himself had been working for fifteen years at "finding some practical solution to the 'Catalogue Problem' which no architect has been able to work out himself." His description of offices overrun with boxes, books and piles of information, and of busy architects with "less and less time" to do "more and more work" still applies today.[23] Although the now multi-volume *Sweets Catalog* has certainly prospered since 1906, becoming an essential tool in virtually every American architectural office, the "catalogue problem" has in no way been solved. Like traffic on the highway system, the flow of building information has only increased in volume and accelerated in speed with each new improvement in information technology.

In 1929, a young Danish architect named Knud Lönberg-Holm sent an article to the *Architectural Record* in which he described the "catalogue problem" as a continuing crisis for the architecture profession, arguing that the solution lay in a radical rethinking of the distribution of information in architecture:

> ... the architect has lost his leadership. From a professional man with a professional ethics he has become a business man subject to the whims of the buyer. The progressive architect acutely realizes that his problem means ultimately the negation of his profession. He has no power to meet his dilemma through his architectural work. As an individual businessman he cannot afford the research work necessary for the proper execution of his ideas; moreover, he is confronted by the gulf which separates him from a client unsympathetic toward an experiment at his expense.[24]

He argued that "collective problems require collective thinking and collective work," and he proposed the invention of an organization that would act as a "clearing house" and "an economically independent research institute," setting standards and organizing information. After a brief stint as a technical editor at *Architectural Record*, he moved in 1932 to found the research office of Sweets Catalog Service. In 1939 he was joined in that effort by the Czech designer Ladislav Sutnar and together they reshaped the look and logic of the catalog, developing the bold graphics and characteristic "S" still used today. Of course Sweets is in no way an economically independent institution. It is produced as a multi-volume bound collection of short catalog sections provided by product manufacturers, whose fees and advertising tie-ins with the *Architectural Record* and *Dodge Construction Reports* directly support Sweets. As a result, most of Lönberg-Holm and Sutnar's work had to be executed indirectly by persuading and teaching manufacturers. They sought to standardize and discipline their advertising inserts, shaping them into documents readily used by busy architects seeking information. In the late 1940s, they formalized their efforts in a pamphlet prepared for product manufacturers and that work was so popular that they brought out an expanded, full color version called *Catalog Design Progress* in 1950. In the introduction they explained that their aim was to produce "dynamic," "living standards" that could keep up with the rapid pace of technological advance:

> Thus with today's industrial development and the concurrent higher standards of industry, corresponding advances must be made in the standards of industrial information itself. The need is not only for more factual information, but for better presentation, with the visual clarity and precision gained through new design techniques. Fundamentally, this means the development of design patterns capable of transmitting a *flow of information...*[25]

Their first section charted the "emergence of new flow patterns" in all aspects of contemporary life — transportation, production, communication — then devoted the body of the book to the visual and structural features with which such information flow patterns should be directed in their catalog. They concluded with a brief theoretical section that offered "flow" as that form of

4.7
Sweet's
Catalog File,
1949.[27]

information that emerges naturally from the functional demands of architectural practice. It was a clever formulation that overcame the form-function opposition that continues to worry modern architects. They explained the emergent condition of flow analogically, by comparison with a variety of other entities newly understood according to the cybernetic concept of *system*: "The flow pattern of any sequence adopts its own form, reflecting function, and its variety of forms may be observed not only in information flow, but in man (the nervous, digestive, and reproductive systems), in industry (production flow), and elsewhere"[26] (figures 4.7 and 4.8).

The management of architectural information by *Sweets Catalog* has continued with the subsequent migration of their catalog information onto compact discs in the 1980s and onto the world wide web in the 1990s, but the original ethic has continued: "Comprehensive information correctly formatted and focused on your customer's needs!"[29] In other words, the flow of product information is always channeled according to a powerful network of interests: according to brand identities and sales relationships, on the one hand, and to the ever-shifting expression of needs, desires and identities, on the other. What Lönberg-Holm's original description did not explain was the degree to which they sought to accelerate that flow of information and increase the pace of industrialization:

> For a continuous advance in production standards there must also be a continuous liquidation of obsolete products, enterprises, and beliefs. This is possible only in an economy where property relations impose no restrictions on the continuous development of new productive forces ... This expansion of social wealth implies increasing industrialization.[30]

4.8
"The flow pattern of any sequence adopts its own form, reflecting function, and its variety of forms may be observed not only in information flow, but in man (the nervous, digestive, and reproductive systems), in industry (production flow), and elsewhere." From *Catalog Design Progress*, 1950.[28]

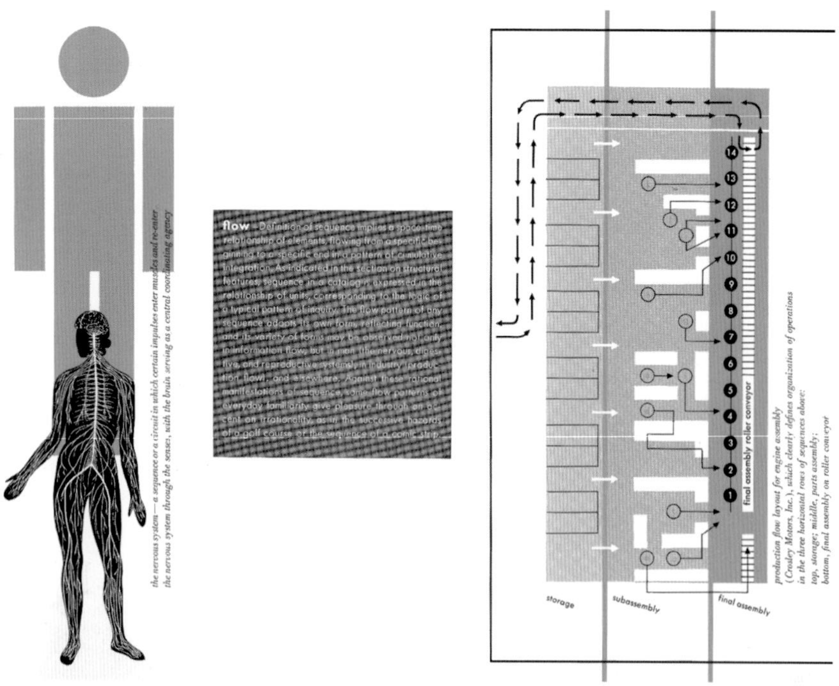

In other words, the system of information flow and industrialized construction has its own momentum fueled by our individual needs, choices and actions. As many critiques have argued, merely fitting better products into normative construction only modulates the effects that industrial development has on the biosphere. To make a difference, it is necessary to understand both the structure and velocities of the flows already in place, and to locate the threshold effects that occur in building.

4.9
Quickborner office communication chart and plan diagram, from *Flexible Verwaltungsbauten.*[32]

	Abteilungsbezeichnung		1	2	3	4	5	6	7	8	9	10	11	
Innerhalb des Geschosses	Geschäftsleitung I	1												
	Geschäftsleitung II	2	12											
	Vertrieb	3	32	–										
	Einkauf	4	20	–	126									
	Rechnungswesen	5	8	–	179	80								
	Export I	6	–	2	9	21	65							
	Export II	7	1	7	32	25	90	6						
	Export III	8	–	8	5	4	15	–	6					
	Inland Leitung	9	16	5	46	45	64	22	44	28				
	Inland I	10	–	–	34	62	93	40	9	5	10			
	Inland II	11	–	–	5	14	4	–	1	–	8	22		
	Summe je Abteilung		89	24	468	398	600	170	223	68	283	275	54	2.628
Untergeschoß	Poststelle, Vervielfält.	12	–	–	24	59	19	13	1	–	51	29	25	
	Lager usw.	13	–	–	7	117	5	10	7	–	48	39	10	
	Firmenfremde	14	2	4	117	96	96	33	15	–	29	109	8	
	Summe 12–14		2	4	148	272	89	56	23	–	128	177	43	952
	Gesamtsumme													3.600

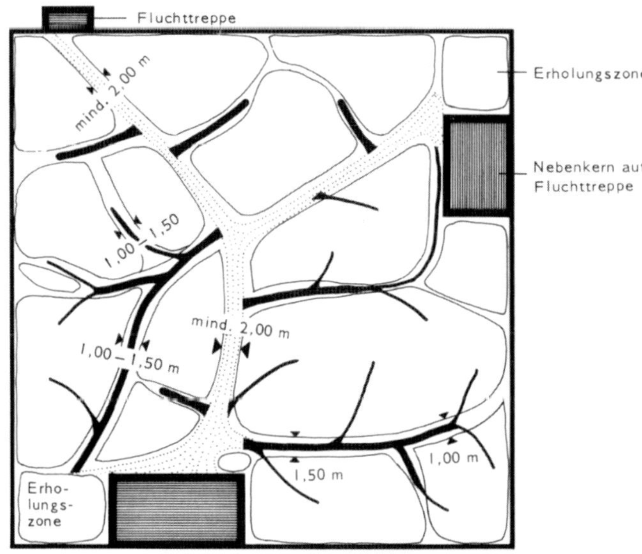

4.10
Quickborner organizational interaction diagram, from *Flexible Verwaltungsbauten.*[33]

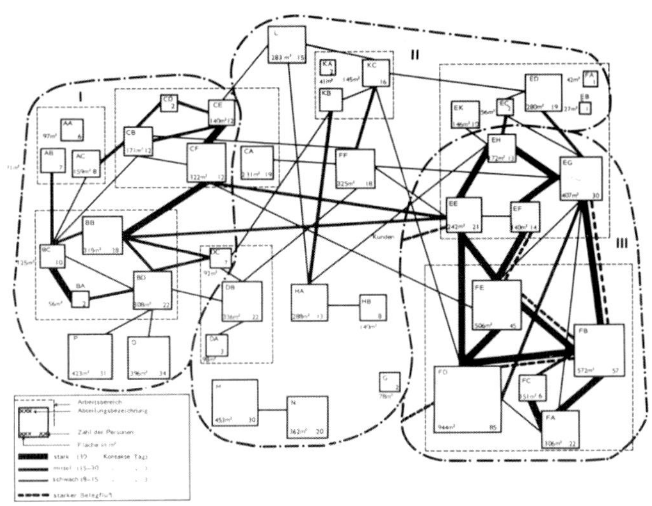

BIOTECHNIQUES: CUBICLES

The acceleration of biotechniques after World War II became evident in research agendas and the rapid development of digital technology. And even as highly rationalized, seemingly mechanized offices were being built across the United States, the German Quickborn management consulting group were quietly inventing a new form of office layout: the *Bürolandschaft* or office landscaping.[31] Based on a rigorous analysis of communication patterns within an office, charted through exhaustive interviewing techniques and diagrams, they dissolved the walls of the office-as-production-line. The analogy to a natural landscape was evident in their pathway diagrams, and in the compelling idea that the form of the office layout was not designed, but emerged from the process of analysis. Their detailed diagrams of communication paths and intensities were the tools that generated the landscape plans, which resembled nothing so much as the meandering "desire paths" that animals, savages and undergraduates chart with their feet (figures 4.9 and 4.10).

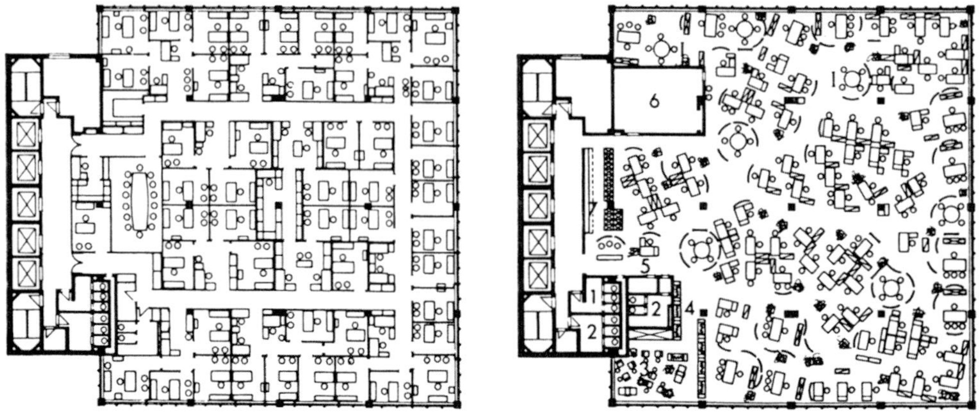

4.11
Application of office landscape techniques to Dupont administrative offices, 1967.[37]

Those ideas were rapidly communicated throughout the planning community and by 1964 the Herman Miller furniture company had formalized them in a revolutionary line of office equipment: Action Office I. Under the guidance of their research director, Robert Probst, they developed the first moveable panels, work surfaces and storage units that came to define the cubicle and made office landscaping possible. By the late 1960s, the effects were visible everywhere and the concept of organic office planning offered a new kind of proportion or regulating system for office layouts:

> The rigid patterns of office layout that had become standard during World War I, assumed the character of time worn tradition by 1960 ... But it failed for precisely that reason. Classical systems are inherently inflexible. Since they embody intellectual-aesthetic ideals of harmony and order, to disrupt any one element is to destroy the whole. Change is inadmissible. When a classical order is imposed upon an organic system — one whose parts are related by functions and processes that are themselves in flux — the result is apparent order and actual chaos. An office is such an organic system. Its organicism, however, is not revealed in those hierarchical charts that bear so curious a relation to feudal concepts of the social orders on earth and in heaven. But, since the actual relations between office personnel defy the caste system codified in charts and embodied in layouts, attitudinal and physical barriers were created that seriously blocked lines of communication.[34]

In close sympathy with structuralist ideas in anthropology and sociology, and exhibitions like *Architecture without Architects* and *Learning from Las Vegas*, the naturalistic forms of *Bürolandschaft* planning offered anti-authorial design strategies that appealed to the generation of 1968.[35] As Francis Duffy reported about his own efforts to spread such ideas, "Anthropology with its rigorous comparative techniques, its search for cross-cultural patterns between artifacts, behaviour, societal norms and their technologies was an obvious model for architectural research. The interrelated three-part model of buildings, people and technology ... was firmly implanted."[36] Even though the organic look of the office landscape passed relatively quickly, the principle of planning around communication, the importance of adaptation and, of course, the cubicle, formed the core of the new biotechniques of the office (figure 4.11).

THRESHOLD EFFECTS: HIGHLY CONDITIONED BUILDINGS

In 1957 the head of the Carrier air conditioning corporation observed that "whenever 20 percent of the office buildings in any one city include air conditioning, the remaining buildings must air-condition to maintain their first class status."[38] That process had apparently taken about ten years, and after the late 1950s it was largely assumed that a high-quality office building in an American city would be conditioned to some degree. The technology had been available for many decades, but it took the particular arms race dynamic of post-war real estate development to change it from a desire to a "need." A similar process had occurred among movie houses in the 1930s, which along with luxury hotels had rapidly adopted air conditioning in the pre-war period once its competitive advantage had been

4.12
A bulky building among other bulky buildings. The classic fully-conditioned building of the late 1950s: Seagram Building, 1958 [40]

4.13
Classic, big-but-thin, unconditioned skyscraper. Sullivan, Wainwright Building, 1891.[42]

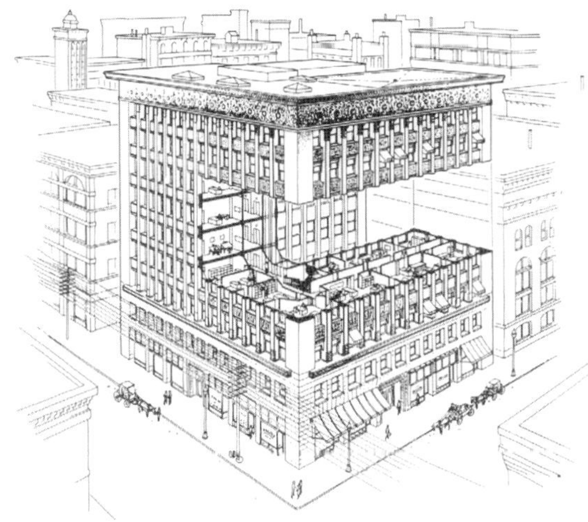

demonstrated.[39] Those examples served to introduce the public to the experience of conditioned air, preparing them for the ever-increasing amounts of conditioning (figure 4.12).

This is one kind of threshold effect that occurs in feedback systems, when an arms race develops between competitors. They rapidly adopt new products, strategies and quite expensive technologies if their customers are free to make other choices. Who would go to a hot movie theater or rent a hot office if a cool one is readily available? And in the process, a new, higher standard emerges and is fixed not only in public desires but in normative construction practices and regulatory codes and standards. At that point, the new standard no longer represents a choice, but a culturally and officially recognized need. It is not easily reversed and can apparently only be altered by a similarly dynamic cultural process. The energy supply crises of the 1970s, for example, temporarily altered thermostat settings and some social habits, but the logic of energy conservation quickly receded when prices dropped.

I do not mean to argue that air conditioning is inherently bad, far from it. The relief from sweating simply feels good, and that is precisely why it becomes such an effective element in competitive situations, leading to a steady escalation of expectations. The problems are twofold. The first are very familiar: greater levels of conditioning produce a whole host of secondary environmental effects through heat island conditions, the use of greater amounts of energy, the release of CFCs, and so on. Many of these are amenable to better design or greater efficiency, and form the basis of most green design strategies, but the second kind of problems are more troublesome. Not only does the escalating aspect of this process establish ever higher standards, requiring ever greater levels of conditioning, but the techniques of conditioning profoundly alter the size and character of the buildings that can succeed in the marketplace.

In other words, once the real-estate process described in 1957 takes place, and conditioning becomes the norm for commercial buildings, then the scale and configuration of those buildings quickly expands so that they have to be conditioned. The dimensions of a commercial building designed without air conditioning are effectively defined by its external skin, meaning that every inhabited workplace has to have ready access to a window for light and air. As a result, even the biggest of the early skyscrapers were made thin by cutbacks, light courts and reentrants. Once the connection to windows is severed by air conditioning and efficient lighting, the buildings are free to grow (out and up) until they encounter other scale limits: circulation, the size of elevators and so on.[41] And like the escalation of comfort standards, this is simultaneously a technical process of conditioning buildings and a cultural one of conditioning the individuals who inhabit them (figure 4.13).

4.14
Classic, fully-conditioned, atrium building with return air circulated through atrium space and plantings. Ford Foundation, 1968.[43]

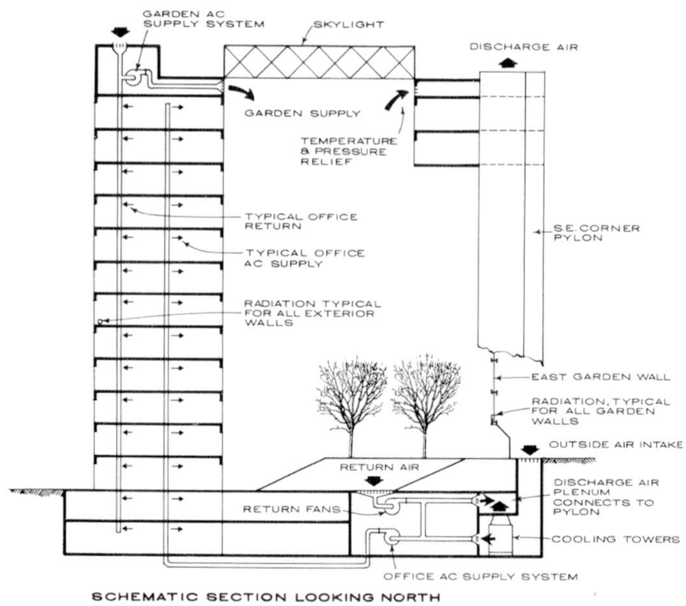

A building's balance-point temperature provides a rough index of when it crosses that threshold, when its spaces are no longer directly connected to the outdoor climate. When a building becomes both sufficiently big and contains a sufficient intensity of internal conditioning and support systems, its balance-point temperature will fall below the average outdoor temperature and it will have to provide cooling for some part of most days of the year (and everyday in their windowless cores). This initiates a fairly simple cascade of effects: first air conditioning and efficient fluorescent lighting make it possible to fill larger interior areas with people and the equipment they use to work, but the people, lights and equipment all produce heat, which requires even more conditioning. As heat removal becomes ever more important, windows are sealed and are designed to exclude as much sunlight as possible, making the interior environment more efficient, but less and less pleasant.

Those two thresholds — higher comfort standards and bigger buildings — were passed for many buildings by 1960, establishing the now familiar norm for commercial and retail construction of highly-conditioned buildings with vast interior spaces. But, of course, that norm has been subject to many criticisms and it has been modified, sometimes radically, in recent decades. Beginning almost immediately in the early 1960s, there were parallel efforts to introduce green plants and natural light into the cores of the newly bulky buildings. The plants initially arrived as part of the office landscape movement (*Bürolandschaft*) and rapidly found a place in the reinvented (and conditioned) atriums of the late 1960s: the Ford Foundation and the Hyatt Regency of 1968 are typically cited as the first fully developed examples. In addition to its pleasant qualities, the atrium was subsequently identified as an energy conservation technique in the late 1970s and 1980s, and become a hallmark of the higher-quality, more efficient office buildings of that period (figure 4.14).

The purpose of this thumbnail history of conditioned buildings is to illustrate the degree to which the environmental thresholds important to green design also involve social and cultural factors, and to explore why they are so resistant to change. A second kind of threshold, one of intensities, is even more critical and difficult to examine because it involves the wholly subjective experience of the bodies being conditioned.

THRESHOLD EFFECTS: BIOTECHNICAL BODIES

The Environmental Protection Agency (EPA) distinguishes "building related illness," which can be attributed to an identifiable cause, from sick building syndrome (SBS) in which "occupants experience acute health and comfort effects that appear to be linked to time spent in a building, but no specific illness or cause can be identified."[44] The inability to diagnose SBS continues, though recent epidemiological studies confirm the correlation between mechanical ventilation rates and reports of SBS symptoms, such as "upper respiratory and mucous membrane symptoms (i.e., irritated eyes, throat, nose, or sinus), and lower respiratory irritation (i.e., difficulty breathing, tight chest, cough, or wheeze)."[45] In this regard, SBS belongs to a broad class of environmental illnesses (EIs), such as multiple chemical sensitivity and Gulf War syndrome, that are widely reported, but that do not fit any biomedical explanation. From one side of the dispute, it is claimed that such syndromes are wholly somatic, learned group expressions of other psychological issues, while on the other side, serious research continues to seek the biomedical causes and etiologies of the distress.[46]

What seems evident in both bodies of research is that the perception of indoor air-quality, of its freshness, is central to the syndrome. As the early ventilation researchers discovered when they first began to investigate ventilation levels in the 1930s, freshness involves both an assessment of the intensity of odors and a judgment about their quality. Like noise, an odor can be pleasant in one situation and offensive or bothersome in another. What this suggests to psychologically oriented researchers is that sensations such as odors can trigger "social psychological processes of contagion, where complaints and symptoms spread from person to person, and convergence, where groups of people develop similar symptoms at about the same time."[47] From the other perspective, the remarkable sensitivity of the nose suggests the possibility of very subtle toxicogenic or allergic processes that have not yet been identified. The statistical correlation between SBS and mechanical ventilation systems, for example, appears to offer evidence of the underlying physical causes related to the rates and processes of ventilation, and has quickly been acted on by design professionals.[48]

I can contribute no new evidence or research that might resolve the biomedical question, but I would argue that as with the previous examples, SBS represents the passing of a critical threshold in the conditioning of buildings, a threshold that is simultaneously physical and social. The previous examples appeared after a certain threshold of scale, after a certain number of buildings were conditioned or after a certain size of building was produced, but SBS and other EIs seem to develop at certain thresholds of intensity. Environmental comfort is defined in these terms, as the intensity of air conditions (temperature, enthalpy, wind, pollution) at which neither our attention nor our coping mechanisms are noticeably required. EI sufferers themselves often explain their symptoms in terms of the cumulative thresholds of toxins or irritants, and they use feedback system theories to explain the disproportionate effects that trace amounts of different substances can cause: "total body load, limbic bundling, and hypersensitivity."[49] For designers, it ultimately makes little difference whether these are medical or somatic explanations, they are the point at which systems designed to provide comfort paradoxically begin to threaten the health of the occupants with the very intensity of their conditioning. As a recent sociological study observed, the accounts of EI sufferers portray "a body that reacts severely to ordinary commercial furniture designed to offer it at least a modicum of rest; a body that responds violently to air passed through conventional heating and cooling systems designed to make it more comfortable… it is as if this body is in protest against the products of modernity and, in its distress, is calling for a radical change in the conventional boundaries between safe and dangerous."[50] Environmental illnesses, like SBS, should remind us that the real object of environmental design is not the efficiency of conditioning, but the state of the bodies that occupy them, whose intimate concerns continue to exceed any performance assessment.

LIVING STANDARDS

I have offered this brief outline of biotechniques to make two very simple points about the conditioning of contemporary buildings. First, environmental conditioning is not just a collection of devices whose performance can be optimized. They are complex systems that operate on buildings and people simultaneously, systems with their own history, trajectory and momentum. Second, there are critical thresholds in the scale, velocity and intensity of that conditioning that radically alter the effects they produce, meaning that more, or even more efficient, conditioning is not always the answer. In that sense, Kiesler was correct, if technological systems self-organize or evolve according to their own performance criteria, then the only useful measure of design is human health. The concept of health is now largely associated with biotechnical medicine, but as SBS illustrates, it still includes social and political forms of coping as well. The best term I can offer as a design guideline for healthy thresholds of conditioning is the "living standard" sought by Lönberg-Holm, a standard that adapts to changing arrangements, and which allows overly conditioned bodies to actively influence their own environments.

To understand what such a living standard might mean for current practice, architects must look beyond the narrowly visual terms which have constrained it. Much of the architectural encounter with environmental conditioning has been devoted to issues of formal expression. The initial opposition between the traditional elements of building — walls, windows and roofs — and the wires, pipes, ducts and devices that invaded them in the late nineteenth and early twentieth centuries, gave way to the "servant" spaces of the Richards Medical Labs, and then to the vigorous display of service elements at the Centre Pompidou. But after fifty years of intensely conditioned buildings, such debates about the expressive role of mechanical equipment seem *passé*.

If we look more closely, however, the history of architectural experimentation reveals a parallel fascination with the symbiotic resolution of buildings and machines. From Le Corbusier's *mur neutralisant* (neutral wall) and Frank Lloyd Wright's radiant floors have sprung an entire ecology of integrated building components, from the "hairy" and "blistered" skin of Roche and Levaux's [Un]plug building (figure 4.15) to the ventilating, double-glass façades of Foster's Commerzbank. Through such biotechnical elements, buildings are not limited to the symbolic expression of cultural ideas, to merely organic forms, but to active demonstrations of the organic themes that lurk within every mechanism.

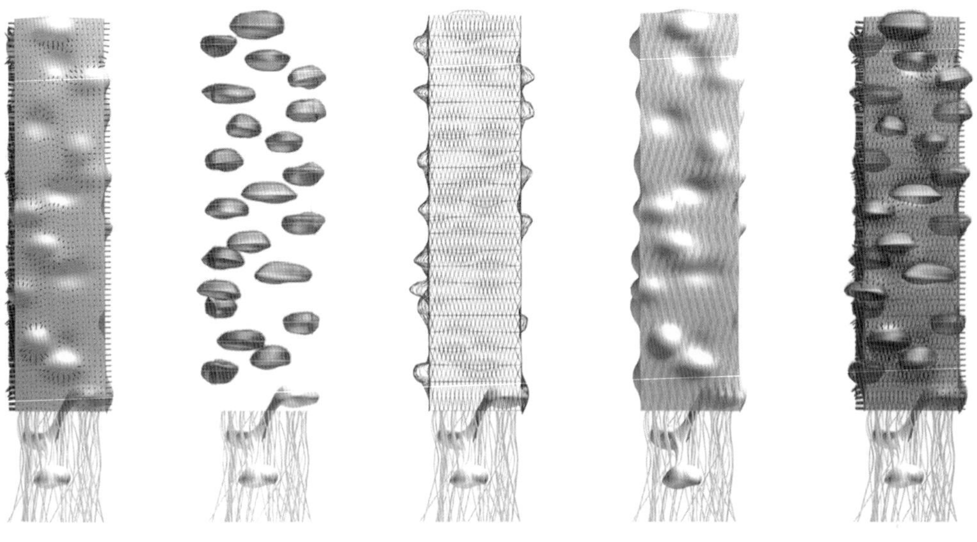

4.15
An "absorbent" concept building, "hairy" with solar collector tubes and "blistered" with photovoltaic cells. Roche and Levaux's [Un]plug Building, 2002.[51]

NOTES

1 "Performance Design" in *Progressive Architecture* 48, August, 1967, pp. 105–153.
2 Frederick Kiesler, "On Correalism and Biotechnique: A Definition and Test of a New Approach to Building Design" in *The Architectural Record,* September, 1939, pp. 60–75.
3 Janine M. Benyus, *Biomimicry: Innovation Inspired by Nature,* New York: Harper Collins, 2002; Karel Honzik, "Biotechnics: Functional Design in the Vegetable World" in *Architectural Review* 81, January, 1937, pp. 21–22; Karel Honzik, "A Note on Biotechnics" in *Circle: International Survey of Constructive Art,* London: Faber and Faber, 1937.
4 Kiesler, op.cit., p. 68.
5 Ibid.
6 Ibid.
7 The mechanical and organic analogies have often been flip sides of the same coin. For an insightful discussion see the chapter on mechanical and organic in Luis Fernandez-Galliano, *Fire and Memory: On Architecture and Energy,* Cambridge: MIT Press, 2000.
8 Dennis L. Meadows *et al., The Limits of Growth : A Report for the Club of Rome's project on the predicament of mankind,* New York: Universe Books, 1972.
9 Jay Wright Forrester, *Industrial Dynamics,* Cambridge, MA: MIT Press, 1961.
10 Jay Wright Forrester, *Urban Dynamics,* Cambridge, MA: MIT Press, 1960.
11 Jay Wright Forrester, *World Dynamics,* Cambridge, MA: Wright-Allen Press, 1971.
12 Forrester, op.cit., *Industrial Dynamics,* p. 191. Fig. 16-4.
13 Forrester, op.cit., *Industrial Dynamics,* pp. 200–201. Fig. 16-7.
14 Forrester, Ibid., *World Dynamics.*
15 Greg Lynn, "Blogs (or Why Tectonics is Square and Topology is Groovy)." In *ANY* 14, 1996, pp. 58–61. "An Advanced Form of Movement" in *After Geometry, Architectural Design* 127, 1997, pp. 54–61.
16 Greg Lynn, *Animate Form,* New York: Princeton Architectural Press, 1999.
17 M. Speaks, "It's Out There ... the Formal Limits of the American Avant-garde" in *Hypersurface Architecture, Architectural Design* 133, 1998, pp. 26–31.
18 A. L. Kroeber and J. Richardson, "Three Centuries of Women's Dress Fashions, a Quantitative Analysis" in *Anthropological Records* 5:2, 1940, pp. 111–153.
19 Ibid.
20 Victor Gruen and Larry Smith, *Shopping Towns USA: The Planning of Shopping Centers,* New York: Reinhold Publishing Corp., 1960.
21 Ibid.
22 S. Lichtenstein, *Editing Architecture: Architectural Record and the Growth of Modern Architecture 1928-1938,* PhD Dissertation, Cornell University, 1990.
23 T. Nolan, "Introduction" and "Sweets" in *Indexed Catalogue of Building Construction,* New York: The Architectural Record Co., 1996.
24 K. Lönberg-Holm, "Architecture in the Industrial Age" in *Arts and Architecture* 84, April, 1967, p. 22
25 K. Lönberg-Holm and L. Sutnar, *Catalog Design Progress,* New York: Sweets Catalog Service, 1950; W. Sanders, "Catalog Design," review of *Catalog Design Progress* in *Architectural Record,* March, 1951, pp. 32 and 35.
26 Ibid.
27 Ibid.
28 Ibid.
29 *Sweets Catalog Services,* 2003: sweets.construction.com
30 K. Lönberg-Holm and C. T. Larson, "The Technician on the Cultural Front" in *Architectural Record,* December, 1936, pp. 472–473.
31 Ottomar Gottschalk, *Flexible Verwaltungsbauten: Panung Funktion Flachen Ausbau Einrichtung Kosten Beispiele,* Quickborn: Verlag Schnelle, 1968.
32 Ibid.
33 Ibid.
34 "From Grid to Growth" in *Progressive Architecture,* November, 1969, p. 100.
35 Bernard Rudofsky, *Architecture without Architects: A Short Introduction to Non-Pedigreed Architecture,* New York: The Museum of Modern Art, 1965; Robert Ventrui, Denise Scott Brown and Steven Izenour, *Learning from Las Vegas,* Cambridge: The MIT Press, 1972.
36 Francis Duffy, *The Changing Workplace,* London: Phaidon Press, 1992, p. 3.
37 Gottschalk, op.cit., p. 221.
38 C. Wampler, "Growth of the Air Conditioning Industry" in *Air Conditioning, Heating, and Ventilating* 54, May, 1957; Gail Cooper, *Air Conditioning America: Engineers and the Controlled Environment, 1900-1960,* Baltimore, MD: Johns Hopkins Press, 1998; Marsha E.Ackermann, *Cool Comfort: America's Romance with Air Conditioning,* Washington, DC: Smithsonian Institution Press, 2002.

39 Gail Cooper, *Air Conditioning America: Engineers and the Controlled Environment, 1900-1960*, Baltimore, MD: Johns Hopkins Press, 1998; Marsha E. Ackermann, *Cool Comfort: America's Romance with Air Conditioning*, Washington, DC: Smithsonian Institution Press, 2002.

40 *Architectural Review*, December, 1958, p. 374.

41 This development has been reviewed by a number of authors: Solar Energy Research Institute (SERI), *The Design of Energy Responsive Commercial Buildings*, New York: Wiley-Interscience, 1985; Dean Hawkes, *The Environmental Tradition: Studies in the Architecture of Environment*, London: E & FN Spon; New York: Chapman & Hall, 1996; A. F. C. Sherratt, *Air Conditioning: Impact on the Built Environment*, London: Hutchinson; New York: Nichols, 1987.

42 Claude L. Robbins, *Daylighting: Design and Analysis*, New York: Van Nostrand Reinhold Company, 1986, p. 5.

43 *Architectural Record*, February, 1968, p. 106. *Progressive Architecture*, February, 1968, p. 94.

44 Environmental Protection Agency, "Indoor Air Facts No. 4 (revised): Sick Building Syndrome (SBS)" in *Office of Air and Radiation, Office of Research and Development, Office of Radiation and Indoor Air 9909J*, April, 1999.

45 C. A. Erdmann, K. C. Steiner and M. G. Apte, "Indoor Carbon Dioxide Concentrations and Sick Building Syndrome Symptoms" in *The Base Study Revisited: Analyses of the 100 Building Dataset*, Indoor Environment Department, Lawrence Berkeley National Laboratory LBNL-49584, 2002.

46 H. Staudenmayer, *Environmental Illness: Myth and Reality*, Boca Raton, FL: Lewis Publishers, 1999.

47 Alan Hedge, "Addressing the Psychological Aspects of Indoor Air Quality" in *Asian Indoor Air Quality Seminar*, Urumqi, China, September 22–23, 1996.

48 O. Seppanen and W. J. Fisk, "Relationship of SBS-Symptoms and Ventilation System Type in Office Buildings" in *Indoor Environment Department*, Lawrence Berkeley National Laboratory LBNL-50046, 2002.

49 Staudenmayer, op.cit.

50 J. Stephen Kroll-Smith and H. Floyd, *Bodies in Protest: Environmental Illness and the Struggle over Medical Knowledge*, New York: New York University Press, 1997.

51 Marc Emery, *Innovations Durables: Une autre architecture francaise = Appropriate Sustainabilities: New ways in French Architecture*, Paris: Ante Prima; Basel: Birkhauser, 2002.

5

PERFORMANCE FORM

THOMAS HERZOG

5

PERFORMANCE FORM

THOMAS HERZOG

Since its founding more than thirty years ago, the architectural practice "Herzog + Partner" has been committed to exercising social responsibility through its projects, while at the same time actively pursuing scientific and technological advances relevant for the environment in multiple ways, such as the potential of harnessing solar energy.

New concepts are developed in collaboration with universities and research institutions, such as the Fraunhofer Association (Fraunhofergesellschaft), the Technical University Munich (Technische Universität München — TUM) and the Center for Applied Energy Research of Bavaria (Zentrum für angewandte Energieforschung Bayern e.V.). Urban planning pilot projects, pioneering buildings, and prototypes of building systems and components are created as a result of the knowledge gained from these interactions. All of the projects incorporate a special demand for aesthetic quality. The form is not predetermined but created, depending on the task, as a result of the design process that we call "performance form."

That working method (the development of "performance form") is the *modus operandi* of our architectural practice. The problem and specific marginal conditions are examined and interpreted systematically. Different alternatives and solutions are formulated on the basis of the philosophy of the architectural practice. In the case of building projects, the local climatic data are recorded in addition to the usual information gathered from investigations, such as the access and positioning of the building in the urban context.

An important factor, which dominates the scope of the practice, is the development of an overall composition that includes building structures as well as the surrounding landscape and public spaces in order to achieve optimal harmony in the architectural design. Using physical models and computer simulations, the effects on the form of the buildings, the positioning, and the possibility of using solar energy for heating purposes, cooling, ventilation, power generation and comfort are investigated. The solutions are found gradually in workshops, together with other members of the design team. The client is also involved in this process, making the decision-making process transparent.

SOLAR DESIGN

In the past, the use of solar energy was seen primarily as a way to reduce conventional heating energy in buildings and to produce hot water. Great advances have been made in both areas through constructional developments. The results can be realized, among other things, by creating large areas of south-facing glazing and closed, heavily insulated north walls, or by zoning in the layout of rooms through the orientation of buildings, thereby ensuring a favorable relationship between the volume and the surface area. In addition, technological advances in the fields of heating and hot-water systems now allow a minimum of 60% of hot water needs in housing to be supplied from solar energy via thermal or storage collectors.

In the 1980s, a disagreement about the use of large areas of south-facing glazing existed. Solar gains were not taken into account when calculating the energy balance of a building; an improved U-value through adequate thermal insulation was seen as the best way to reduce fossil fuel consumption. This came to be recognized as a monocausal view, as it largely ignored the fact that buildings are highly complex organisms, functionally, technically and aesthetically.

Even in moderate climates, increased insulation can lead to cooling problems in the summer, especially in administration buildings. On average, less than 10% of the overall energy consumption of office buildings is attributable to heating. The energy needs for cooling, on the other hand, represents between 10 and 20%; and cooling requires roughly three times as much primary energy per KW hour as heating. Today, components with variable g-values (the total energy-transmittance value through a material) have become a useful tool in the construction of external walls, allowing buildings to respond differently to changes in the weather. New solar cooling systems are also a promising area of development. Maximum energy power is available for this purpose when it is most needed. If one reduces the areas of glazing, however, less daylight will enter a building, thus the proportion of artificial lighting that is required will be increased. Research has shown that in Europe the average proportion of energy required for artificial lighting in administration buildings is in the range of 30%.[1]

Computer tools have long been available for simulations and data processing in various fields of construction, such as the thermal or lighting balance of buildings. Many new components are being developed or are already being tested in these areas. They include vacuum thermal insulation, new types of glass that can react to the changing needs of insulation, such as switchable glass with an inert-gas filling and with U-values of around 1.0, electrochromic and thermotropic glass, ventilation components with a solar preheating facility, and many other innovations.

Environmental forms of energy can be used for a variety of purposes — for natural lighting, ventilation, heating and the generation of electricity via photovoltaic systems. As the conditions change according to the season, the time of day, the weather and the type and duration of use, conflicting needs can arise. One might expect a solution to these problems could be provided by the so-called "intelligent building" systems. These systems respond to changing conditions and can be used to control different functions, such as heat generation, distribution and output, the operation of sunscreen blinds, co-ordination of daylighting and complementary artificial lighting, ventilation flaps, and humidifiers. Automatic controls of this kind are largely powered electrically and are a common feature in our lives today. Certain reservations, however, can be attributed to extensive automation. These include the vulnerability of the technical systems themselves, higher construction costs, and a lack of awareness of cause and effect on the part of users. It is important to understand the phenomena one activates. Electronic systems in buildings must, therefore, serve the sense of orientation of the people who use them. Human beings should experience their environment — including the artificially created elements — with all of their senses, in order not to suffer from a process of mental and spiritual atrophy.

URBAN CONSIDERATIONS

In the first half of the 1960s, it became clear that the segregation of urban functions into housing, production and leisure zones led to a loss of quality in modern cities. At that time the interest was concentrated on the recovery of complexity. "Concentration, Interconnections, Urbanity" was the title of a seminar held in 1964 at the Department for Urban Planning of the TUM. Younger architects pinned their hopes to large-scale variable structures of the kind that Yona Friedmann had developed for Paris, Eckhard Schulze-Fielitz for the Ruhr area, Kisho Kurokawa for Tokyo, and Kenzo Tange for the Tokyo Bay project. In these schemes, proposals were made for handling massive increases in traffic, especially automobile traffic. The volume of traffic itself was not in question. Today we know that transport accounts for roughly a quarter of all fossil energy consumption, in addition to being responsible for a number of negative side-effects. It is necessary, therefore, not only to replace fossil fuels and reduce the volume of private traffic, but to reconsider the causes. Linking urban functions can be a solution because the spatial separation of industrial uses from housing is not as necessary today as it was in the previous century. In concrete terms, this means that a mixing of functions should be reintroduced. Urban structures that are more neutral in their internal circulation and in their use of space are needed. An additional improvement would be to increase building density wherever possible.

Only in neighborhoods with sufficient purchasing power, and where the facilities serving daily needs are accessible on foot, is an effective mixture of functions viable; only under these circumstances can vehicular traffic be reduced significantly. This would also lead to a reduction of land use and infrastructure costs. The ideal of living in a green suburbia on the edge of the city may convey a feeling of proximity to nature, and thus possess a spurious ecological appeal. However, it causes dependency on transport and an increase in fuel consumption and environmental pollution, as well as disastrous social consequences. The lack of residential space with urban quality in our city centers is the cause of the enormous numbers of commuters. Buildings also provide too little scope for change. Millions of square

5.1
Site plan, Guest Building for the Youth Education Center, Windberg, Niederbayen, Germany (1987–91), architect Thomas Herzog with Peter Bonfig and Walter Götz.

meters of developed space remain unused, preventing access to energy and material resources. In addition, we allow ourselves the luxury of an average of 40 m^2 of living area per person in Germany, in comparison to 15 m^2 in Italy and much less in Moscow. Anyone who fears that higher densities in cities would lead to the insalubrious conditions of the past overlooks the improved material state of our society.[2]

GUEST BUILDING FOR THE YOUTH EDUCATION CENTER[3]

The Premonstratensian monks of Windberg Monastery run a youth education center for which they commissioned a new dormitory block. The nature and duration of use of the various groups of rooms played an important role in determining the energy concept for the building. The rooms that are used for longer periods of the day are separated from those that are used briefly. The two sections of the building were also constructed with different materials (figures 5.1 and 5.2).

5.2
Guest Building, Windberg: upper, ground and lower ground floor plans.

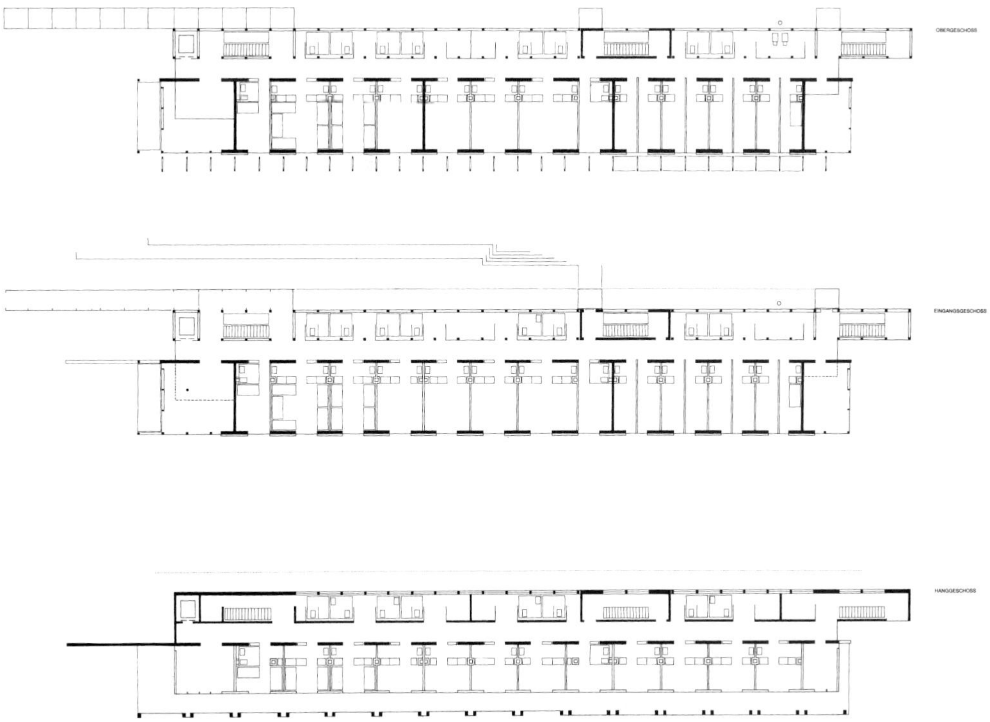

5.3
Guest Building, Windberg: southern façade.

The southern tract houses the lounge areas and the bedrooms, which can be divided in different ways. The rooms on this face are more attractive, with an open view of the landscape through broad areas of glazing (figures 5.3 and 5.4). A direct exploitation of solar energy and daylight was also possible for heating and lighting the spaces on this side, which are used for longer periods. To reduce temperature extremes and to permit storage of thermal energy, this tract was constructed with heavy, thermally sluggish materials. Opaque areas of the south-facing external walls were also clad with translucent thermal insulation (figure 5.5). This allows solar radiation to pass through it, but minimizes thermal losses. The south-facing external wall is thus heated up during the day and passes on the thermal energy to the internal spaces after an interval of five to six hours, beginning in the early evening (figure 5.6). In other words, during the night, when external temperatures are at their lowest, the outer wall functions as an inward-facing solar heating area (figure 5.8). During the summer months, overheating is prevented by the broad roof projection and external louvered blinds.

5.4
Guest Building, Windberg: eastern façade.

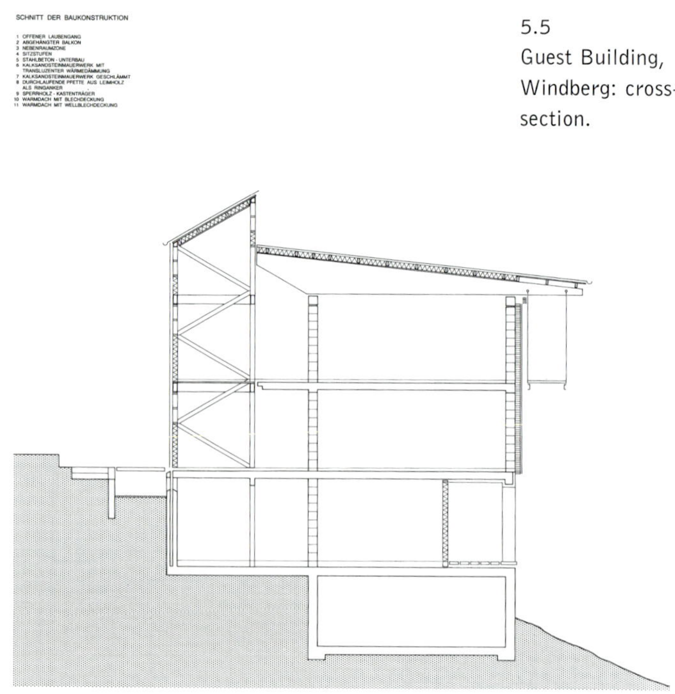

5.5
Guest Building, Windberg: cross-section.

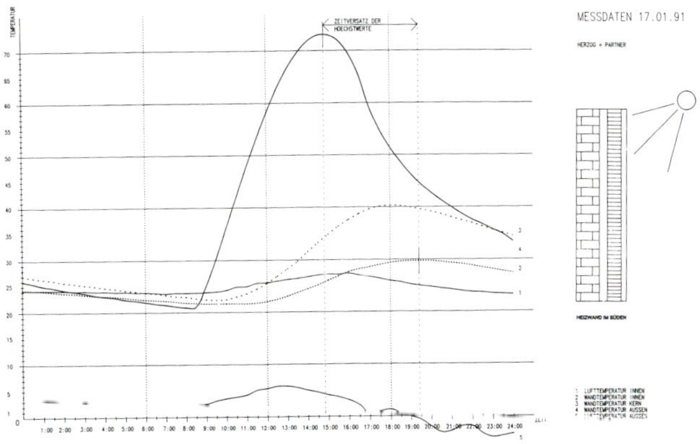

5.6
Guest Building, Windberg: temperature curves in south-facing external wall on a clear January day.

5.7
Guest Building, Windberg: northern façade.

The northern tract contains the sanitary facilities and storage spaces, as well as the circulation route through the building. These spaces are distinguished by the fact that they have a generally lower average temperature level, since they are used for only short periods. In the shower rooms, for example, a higher temperature level is required for only two to three hours a day. This tract was, therefore, equipped with a quickly functioning warm-air heating system. To minimize heat losses through ventilation, a heat-recovery unit was installed in the attic space. The hot-water supply is provided largely from solar energy by means of vacuum-tube collectors on the roof.

Part of the teaching program of the youth education center is to make the functioning of the building comprehensible to young guests. This is accomplished by providing them with insights into the use of environmentally sustainable forms of energy through "passive" and "active" constructional systems and through the mechanical installations that play a role in the energy balance. The architectural effect of the newly developed south-facing heating wall is immediately visible in the façade and is tangible internally. The service runs, solar storage units and collectors are exposed to view, and a display panel which is installed in the entrance area shows changes in temperature levels.

DESIGN CENTER[4]

The congress and exhibition hall in Linz marks a new interpretation of the concept of early historical examples of large glazed halls. These include the Crystal Palace in London (1851–1936) and the Glaspalast in Munich (1854–1931), which provided effective protection against the elements and a hitherto unknown internal light quality. From the outset, one of the goals in planning the Design Center in Linz, Austria (1989–93), was to reduce the inner volume of air to a minimum. The internal height of the spaces was limited to 12 m, and since this clear height was not required everywhere in the hall, the roof structure was designed in a flat arched form with a glazed covering (figures 5.8 and 5.9).

The steel girders forming the load-bearing roof structure span a distance of 76 m and cover an area of 16,800 m^2. To ensure maximum flexibility of use, all exhibition and congress spaces (with a capacity of 650 and 1,200 people) adjoin a common foyer (figures 5.10 and 5.11). The points of access are laid out in such a way that visitors to concurrent events do not interact. Continuous longitudinal access routes along both sides allow the various halls and the gallery space to be combined. The ancillary zones are also laid out in linear form. Since the partitions in these zones can be moved, the spaces remain flexible for the changing uses (figures 5.12 and 5.13).

In developing a natural lighting concept for the building, the challenge was to achieve excellent light quality in exhibition areas without having to make sacrifices in the indoor climate and without giving rise to excessive energy consumption. In collaboration with the Bartenbach LichtLabor, a new kind of building element was developed for the light-transmitting roof (figure 5.14). A plastic grid integrated in roof panels with a complex performance allows indirect luminous radiation from the northern hemisphere of the sky to enter the building, while direct sunlight is screened off (figure 5.15). In this way, excessive heat gains are avoided in internal spaces in the summer. Just 16 mm deep, the retro-reflecting grid, thinly coated with pure aluminum, was inserted into the cavity between the panes of double-glazing over the roof (figure 5.15).

5.8
Sprawling Design Center, Linz, Austria (1989–93), architect Thomas Herzog with Hanns Jörg Schrade and Heinz Stögmüller.

5.9
Design Center: model.

5.10
Design Center: exhibition hall with the ventilation system.

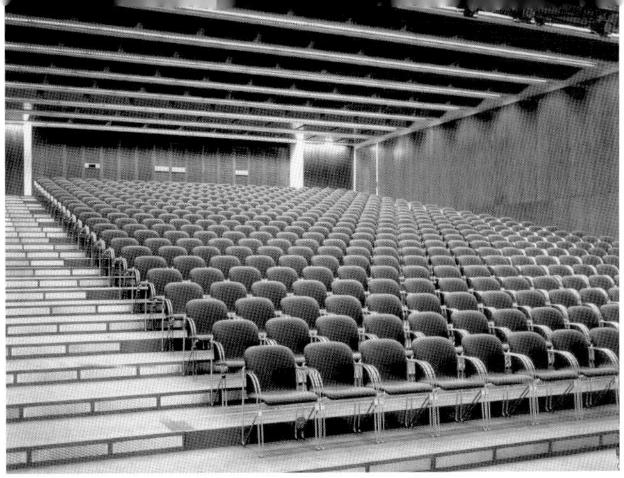

5.11
Design Center:
congress hall.

5.12
Design Center: steel
girders framing the
roof structure.

5.13
Design Center: end
wall with the clay-
tile façade.

5.14
Design Center:
simulation of the
light-transmitting
roof.

5.15
Design Center:
entrance hall.

5.16
Design Center: long face of the hall with "Venturi" capping to assist natural ventilation.

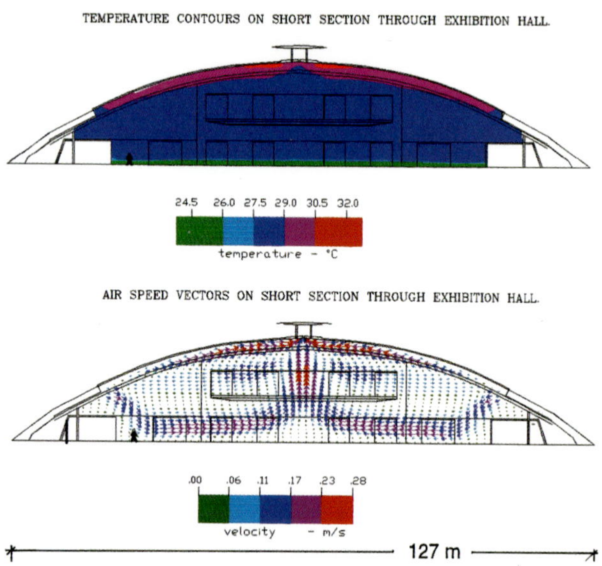

5.17
Design Center: simulations of temperature curves and airflow patterns.

The geometry for cutting the grid was determined by computer programs and had to take account of the following factors: the angle of elevation and the azimuth angle of the sun at various seasons, the exposure and orientation of the building, and the slope of the roof. Thermally separated steel sections help to reduce heat losses through the building envelope.

In addition to thermal and daylighting aspects, a special challenge was posed by the need to guarantee an adequate air change in this flat, deep building. Fresh air enters via floor inlets and ventilation flaps at the sides of the hall. The warmed, used internal air rises to the top of the building as a result of thermal buoyancy. During the heating period, the air is then borne by large ducts to a heat recovery plant. During the rest of the year, the exhaust air escapes from the building at the crest of the roof via a large, continuous opening that is fitted with closable louver flaps. To guarantee the extraction of the vitiated air under unfavorable air-pressure conditions, a "spoiler" capping was developed and assembled over the crown of the roof (figure 5.16). This 7 m wide element has a convex underside and exploits the "Venturi effect" to support the extraction of air from the building (figure 5.17). The final form of this element was determined in wind-tunnel tests (figures 5.18 and 5.19). In light of these developments, one sees that building envelopes are subject to changes in their technical functioning and construction when, in addition to performing their traditional protective role, they are required to control indoor temperatures and the ingress of daylight.

5.18
Design Center: simulations using the wind tunnel.

5.19
Design Center: simulations using the wind tunnel.

5.20
Aerial view, Deutsche Messe AG (DMAG) Administration Building, Hanover, Germany (1997–99), architect Thomas Herzog and Hanns Jörg Schrade with Roland Schneider.

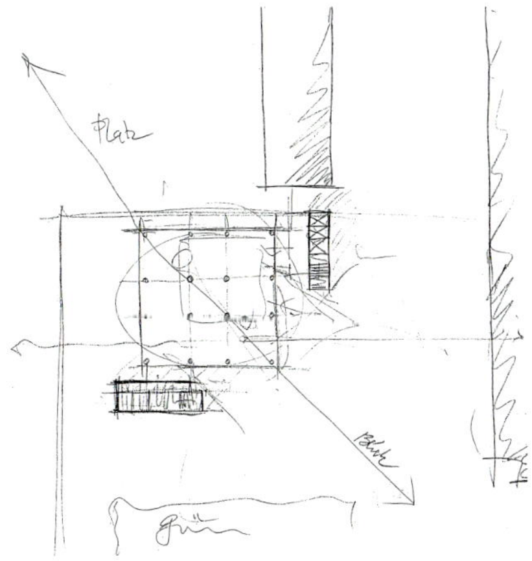

5.21
DMAG: conceptual sketch.

5.22
DMAG.

DEUTSCHE MESSE AG (DMAG) ADMINISTRATION BUILDING[5]

As a building type, high-rise buildings are not usually regarded as compatible with the conservation of resources. This project proved it was possible to design a "sustainable" building that refutes this opinion. This is accomplished through a new interpretation of spatial and functional concepts, by co-ordinating the form of construction with the energy concept, applying sound principles of building physics, and exploiting locally available forms of environmental energy.

Taking account of environmentally relevant issues, high quality in the workplace and flexibility of use were the main criteria in ensuring that the building could adapt to changing working needs over time (figures 5.20–22).

The layout is articulated into a central working area of 24 x 24 m in plan and two tower-shaped access cores, containing ancillary spaces, which are offset to the sides. This allows great flexibility in the use of the building, which is twenty stories high. Above the three-story entrance hall are fourteen floors that are used exclusively for offices. At the top of the building are conference and discussion spaces, as well as a story occupied by the company management. The individual floors can be divided into open-plan, combination or single-unit offices as required, whereby a similar quality is guaranteed for every workplace (figures 5.23–28).

All users can enjoy natural ventilation by opening the sliding casement doors to the intermediate space between the two skins of the façade. When the casements are closed, fresh air is supplied via inlets from the ventilation ducts incorporated in the inner façade skin. Vitiated air is extracted from the offices by means of thermal uplift of the warmed air in the internal spaces and is channeled through a central duct system, with vertical shafts leading

5.23
DMAG: ground floor.

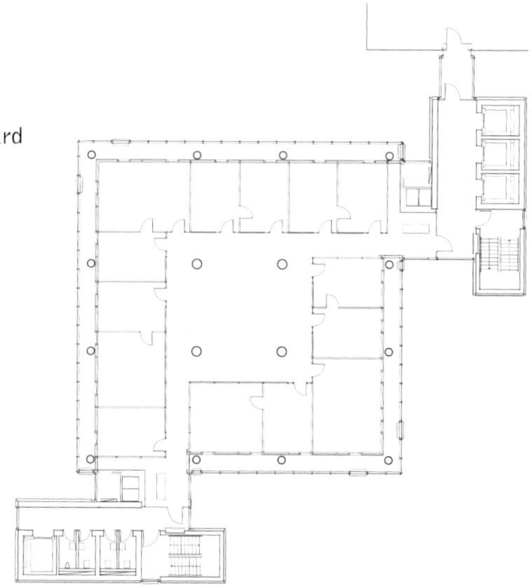

5.24
DMAG: standard floor plan.

5.25
DMAG: Hermes Lounge.

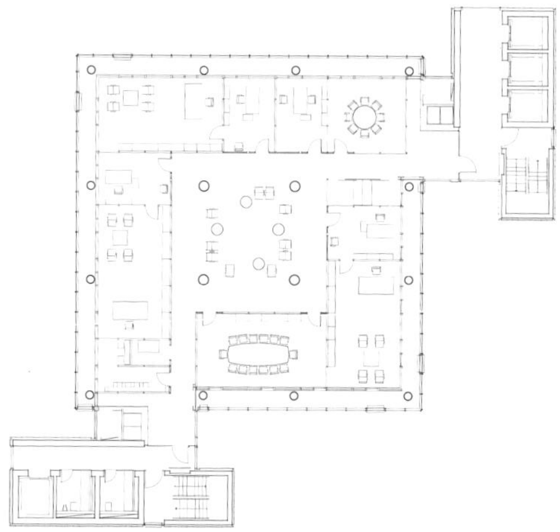

5.26
DMAG: board members floor plan.

up to a rotary heat-exchange unit. In winter, this allows 85% of the thermal energy contained in the extracted air to be used for preheating the fresh-air intake.

The integration of thermal storage mass into the overall concept was of great importance in ensuring an efficient use of energy and a high degree of internal comfort. The heating and cooling system laid in the monolithic screeds allows the thermal environment to be controlled at a low temperature level. By storing heating or cooling energy in the thermo-active floor slabs, which is released at a later time, it is also possible to reduce temperature extremes, thus ensuring a balanced indoor climate and agreeable surface temperatures on the space-enclosing elements.

A distinguishing feature of this structure is the co-ordination of the various building subsystems within an overall concept. In this way, it was possible to guarantee a high level of comfort with low energy consumption, and to harness sun and wind energy to control the indoor thermal environment and ventilation. A ventilation tower rises by about 30 m above the northern access core. The exploitation of thermal uplift is an important aspect of the natural air-supply and extract system for the entire building (figures 5.29 and 5.30).

The load-bearing structure consists of a reinforced concrete skeleton frame with in-situ concrete floors. The building is braced by the two access towers, which, in conjunction with the floor slabs, form a stable structural system. The access towers are clad with the Moeding façade system, a rear-ventilated clay-tile form of construction suspended from the main structure.

The double-skin glazed façade to the office areas (figure 5.31) offers several advantages. The glazed outer skin acts as a screen against high-speed winds, thereby allowing natural ventilation (figure 5.32). Sunshading can be installed in a simple form behind the outer façade layer, where it is protected from the elements and is easily accessible for maintenance and cleaning (figure 5.33). The buffer effect created by the corridor space between the two façade skins (figure 5.34), and the high resistance to

5.27
DMAG:
glazing.

5.28
DMAG:
terrace.

5.29
DMAG: the southern and eastern façades.

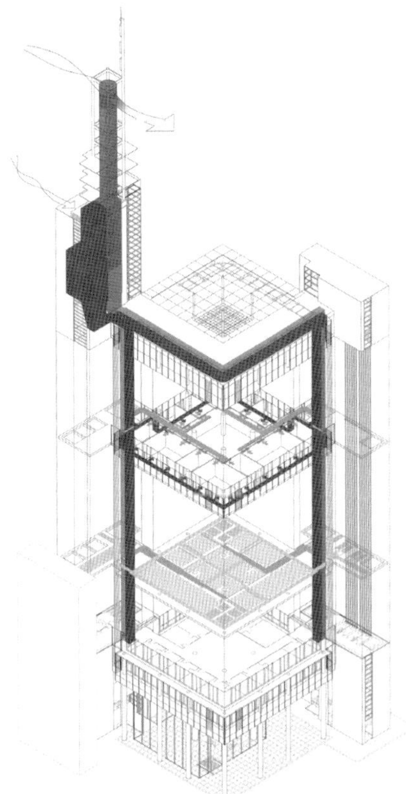

5.30
DMAG: the northern and western façades.

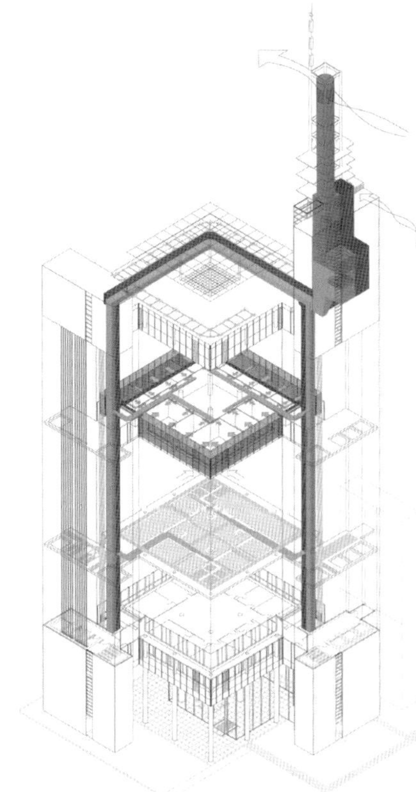

5.31
DMAG:
double-skin
façade with
fixed glazing.

5.32
DMAG: cross-section through the double-skin façade with external fixed glazing.

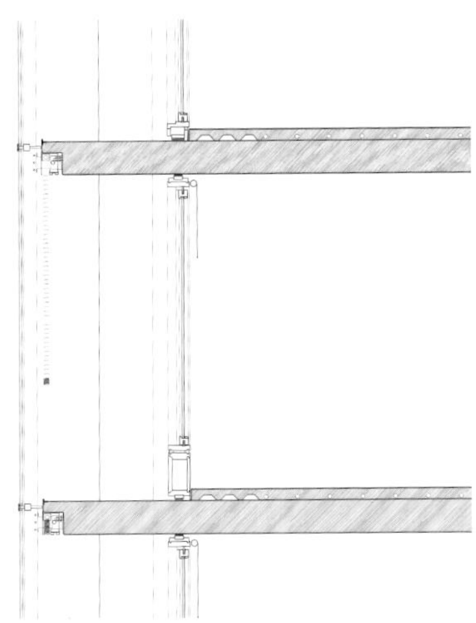

5.33
DMAG: cross-section through the double-skin façade with external ventilation flaps.

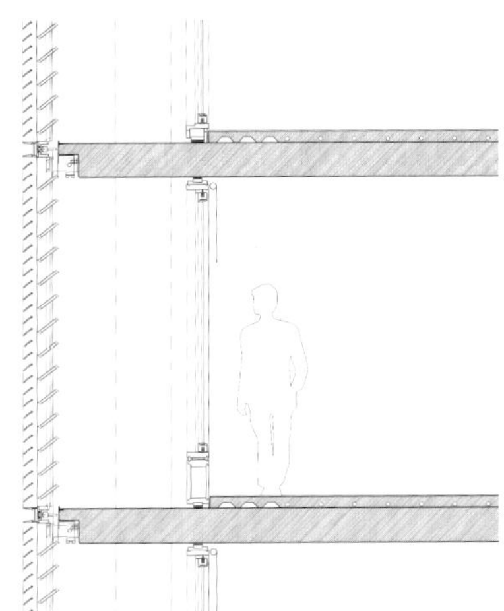

thermal transmission provided by the two layers of glazing, help to reduce the effects of insulation near the surface of the inner façade and increase the sense of comfort in the rooms. The cantilevered reinforced concrete floors and the fire protection they provide allow a form of façade construction with story-high glazed elements. This, in turn, facilitates a maximum exploitation of daylight and creates an ample sense of internal space. The cantilevered section of the floor slab does not have to be thermally separated from the main area as a result of the use of insulating double-glazing in the outer façade skin. In addition, it was possible to locate the load-bearing columns in the façade intermediate space where they do not obstruct the functional floor area.

5.34
DMAG: intermediate space between façade skins.

CONCLUSIONS

Architecture is one of the few professions with a truly comprehensive character. Architects have to take a holistic approach to complex systems. It seems likely that the methods and patterns of work we know today will be subject to fundamental changes in the coming decades. This applies, in particular, to the nature of co-operation between architects and specialists in other fields. A large part of the problem we face today in respect to the natural resources is attributable to one-sided processes of optimization, in which insufficient attention is paid to potential side-effects. Success comes more easily by ignoring disturbing secondary issues than by seeking a balance between the various factors involved. If we are to achieve this balance between technical processes, the harmony of nature and a sense of social responsibility, we must ensure that a long-term, mutual responsibility for the public welfare plays a greater role in life.

A society that wishes to develop along humane lines must be more than the sum of its individuals pursuing their own interests. A holistic approach to problems can be achieved only if we succeed in intensifying interdisciplinary thinking in collaboration among the arts, natural and social sciences, engineering and economics, and if environmental design is understood as a complex central discipline.

NOTES

1 According to investigations carried out by Professor Nick Baker, presented in a lecture in Cambridge, England.
2 Adapted from a revised version of a lecture given in the German Architectural Museum in Frankfurt on Main on the occasion of the exhibition "The Ecological Challenge" in February 1997.
3 Windberg, Niederbayen, Germany (1987–91), architect Thomas Herzog with Peter Bonfig and Walter Götz.
4 Linz, Austria (1989–93), architect Thomas Herzog with Hanns Jörg Schrade and Heinz Stögmüller.
5 Hanover, Germany (1997–99), architects Herzog + Partner, Thomas Herzog and Hanns Jörg Schrade with Roland Schneider.

6

PERFORMANCE SIMULATION: RESEARCH AND TOOLS

ALI M. MALKAWI

6

PERFORMANCE SIMULATION: RESEARCH AND TOOLS

ALI M. MALKAWI

Simulating building performance requires specialized expertise that targets the design, engineering, construction, operation and management of buildings. It draws its resources from many diverse disciplines, including physics, mathematics, material science and human behavior. Its intention is to predict the behavior of a building from conception to demolition.

Most of the fundamental work on building performance simulation algorithms and predictions was developed a few decades ago. Simulation tools have become widely available and this has had a measurable influence on the way in which buildings are designed, analyzed and constructed. However, building simulation continues to evolve and the outcomes of research in this area are being incorporated into the design and the construction of most buildings, and they are evident in the recent advances in building simulation environments. This progress was driven primarily by funded research efforts (academic and governmental) and commercial kernel encapsulations, with both sectors benefiting from advances in computation, such as new programming paradigms and the increased power of computing and the Internet. Funded research involved the development of simulation throughout the life cycle of the building. Efforts in commercial development have been primarily purpose driven, taking advantage of the maturing algorithms which are able to resolve particular issues of building performance. These environments focus on providing user-friendly interfaces while allowing for flexibility in modeling and accuracy.

Advances in building simulation environments have been focused on two areas: the structural framework of these environments and the activities they support. The structure of the simulation involves algorithm developments, data management and interfacing. Simulation algorithms have historically been designed to predict answers to domain-based questions, such as lighting, thermal or structural problems. Each of these domains has sub-problems that must be modeled and simulated differently. These maturing algorithms are under rapid development in the areas of code validation, uncertainties and efficiency of representations.

To shift the conventional use of such tools from analysis only to analysis and synthesis, a renewed research into utilizing advances in optimization is underway. This research stems from the idea that digital simulation tools can be used to support performance-driven design using optimization and partial automation. Current research promises to take advantage of advances in visualization and human computer interaction developments. It is responding to the current needs of the design team, which are far from being fulfilled. Methods to assist in coupling and data management between algorithms to increase their prediction accuracy are also being explored. In addition, frameworks and standards are being developed to facilitate their integration with other environments in order to support the design analysis team's activities. The following sections in this chapter provide a summary of these developments and advances, and discuss the challenges that exist.[1]

SINGLE DOMAIN TOOLS

Many tools have been developed to predict the performance of the building in areas such as thermal flows, lighting, acoustics, structures, etc. Because different problems require different simulation algorithms, a variety of computational simulation tools exist. The United States Department of Energy (DOE) maintains an online directory of energy-related tools for buildings.[2] These tools vary from simple, approximate performance tools to the very precise. The website demonstrates the diversity of the tools for solving problems of energy-related issues in buildings.

Although refinements and updates have been made to many of the single domain environmental simulation tools, the two main developments within the past decade are the introduction of EnergyPlus™ and a surge in the use of Computational Fluid Dynamics (CFD).

In 1995, the DOE introduced EnergyPlus™ to replace DOE-2 (sponsored by the DOE) and BLAST (sponsored by the US Department of Defense), which is based on the best capabilities of both tools. It launched its first release in 2001 with a promise of continued support and development. EnergyPlus™ is a simulation engine without

6.1
SmokeView — burn room project showing 3D smoke visualization.[11]

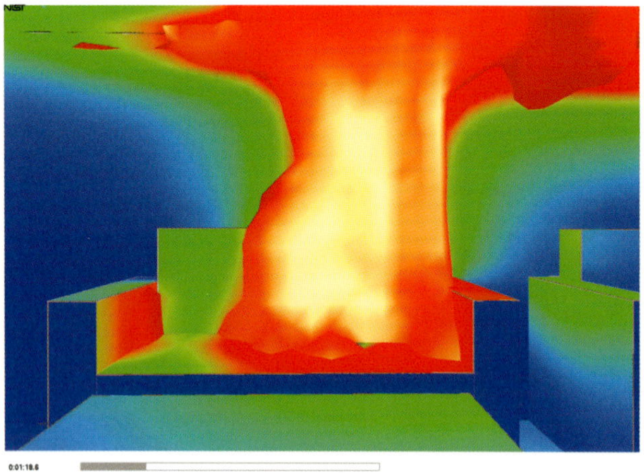

a user-friendly interface that simulates heat and mass energy flows throughout a building. The concept is to provide a rigorous objected-oriented environment that will attract third-party developers to create user interfaces and modules, and that can be linked to other programs through data and object interaction, as described later in this chapter.[3] This concept marks a new era in the way simulation engines will be developed and supported. It also marked a shift from approximate-based simulation and simplified interface development for the non-expert to the support of rigorous engines that can be utilized in frameworks and environments to aid in the building design for both the non-expert as well as the expert user.

As computing power became less expensive, the use of CFD (after evolving for several decades) began to increase. CFD applies numerical techniques to solve the Navier-Stokes equations for fluid fields and it provides an approach to solve the conservation equations for mass, momentum and thermal energy. CFD has been used in many applications in relation to buildings. These include natural ventilation design,[4] building material emissions for indoor air-quality assessment,[5] and complex flows of fire and smoke in buildings.[6] Fire simulation using CFD is exemplified by the National Institute of Standards and Technology's open source Fire Dynamics Simulator (FDS) (figure 6.1).[7] FDS predicts the smoke and hot air flow of a fire by using Large Eddy Simulation. Using different parameters, CFD is also used in noise prediction in relation to ducting in buildings.[8] Other applications are more complicated and may integrate alternative building simulation models, such as EnergyPlus™ described earlier. Although it has been widely used, the technology of CFD is still under development as applied to the solution methods used.[9] Studies illustrate the importance of following validation procedures to ensure the accuracy of the CFD results.[10] It is important to highlight that CFD remains an expert tool and its use requires knowledge of fluid mechanics to set up the simulation model, populate its boundary conditions and interpret the results (figures 6.2–6.4).

COLLABORATIVE ENVIRONMENTS AND INTEROPERABILITY

In order to overcome the drawback of needing different simulation engines to predict different performance aspects, integration between algorithms provides a means to overcome some of these limitations and increases the efficiency and prediction accuracy of performance tools. Research in this area has predominantly involved issues of inter-model coupling based on engine and equation integration.[12] On the other hand, as the environments shifted from procedural to object oriented, simulation-based environments followed this shift. Object-oriented code that supports modularity and inheritance allowed simulation to be more flexible and expandable. This shift made it "technically" possible to begin work on encapsulating shared and distributed simulations. It also made possible the development of environments and frameworks that can achieve design analysis integration based on semantic representations which support object and data interaction. This includes custom-driven object integration developments, the data and object interoperability development, as well as the research in process-driven integration.

Custom-driven data integration attempts to integrate multi-simulation engines into one system that provides interoperability between the various tools by using shared custom data objects. Examples of such

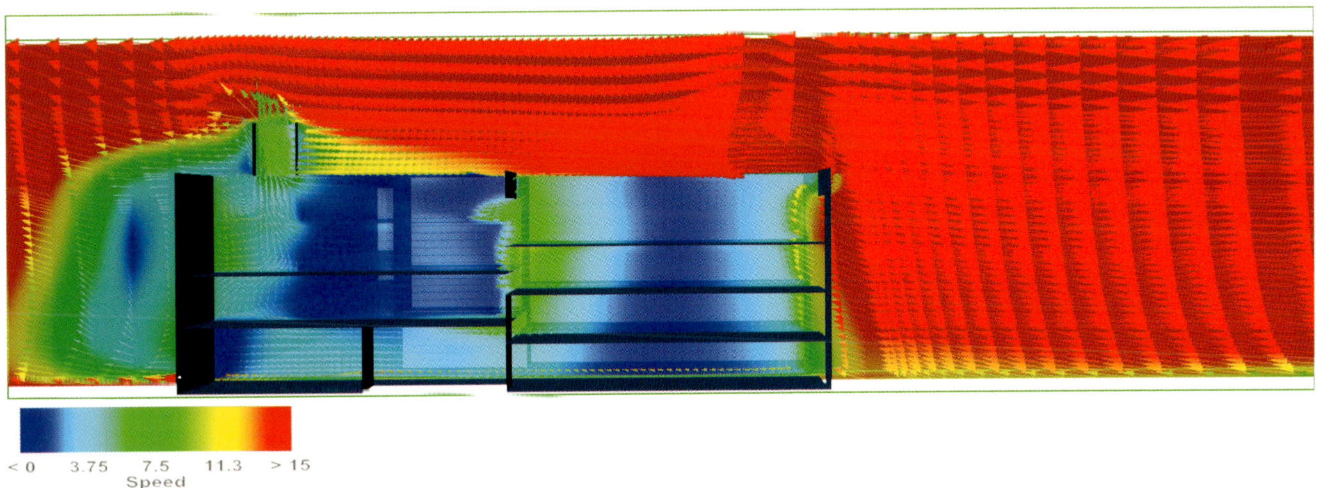

6.2
CFD simulation and analysis for the PennDesign building during the schematic design phase.

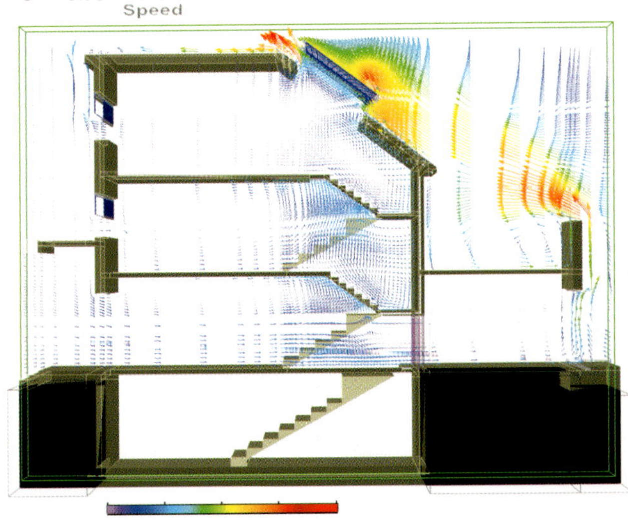

6.3
CFD simulation and analysis of Civic House, University of Pennsylvania, for renovation and retrofitting purposes.

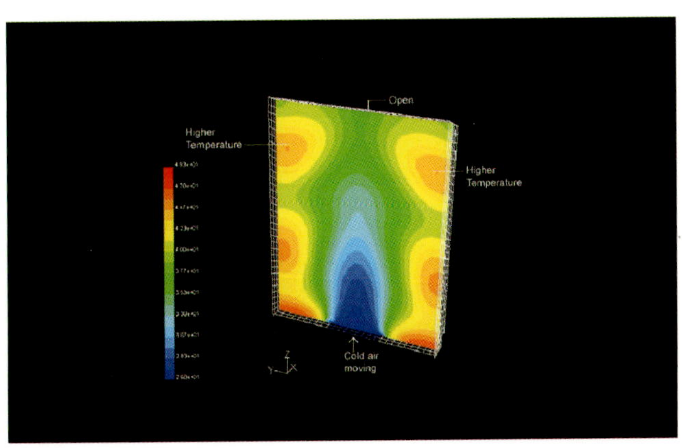

6.4
Study of double-skin façade (New York Police Station) using CFD simulation.

development are the Building Design Advisor (BDA)[13] and SEMPER environments.[14] Although SEMPER differs from the BDA in the way it represents the design-analysis activities and its internal interfaces, both use "custom" building model representations based on object-oriented technology.[15]

As interoperability among different software became a necessity, a collective effort by industry, governmental and research organizations to establish data exchange standards for the building industry attributed to a surge in object- and data-driven integration. This approach targets data sharing among different software applications, including simulations, to achieve software interoperability. Data sharing is achieved by mapping the relevant data within each program to a generic common data model that contains information required by all other programs. The development of this common data model is currently the mission of the International Alliance of Interoperability (IAI), which was formed in 1994 and built on the foundation of previous research projects in data exchange for integrated building models and standards in related areas.[16] The view is that this universal framework and specifications for the common data model — called the Industry Foundation Classes (IFC) — for the architectural, engineering, construction and facilities management industries can facilitate interoperability between the existing and future

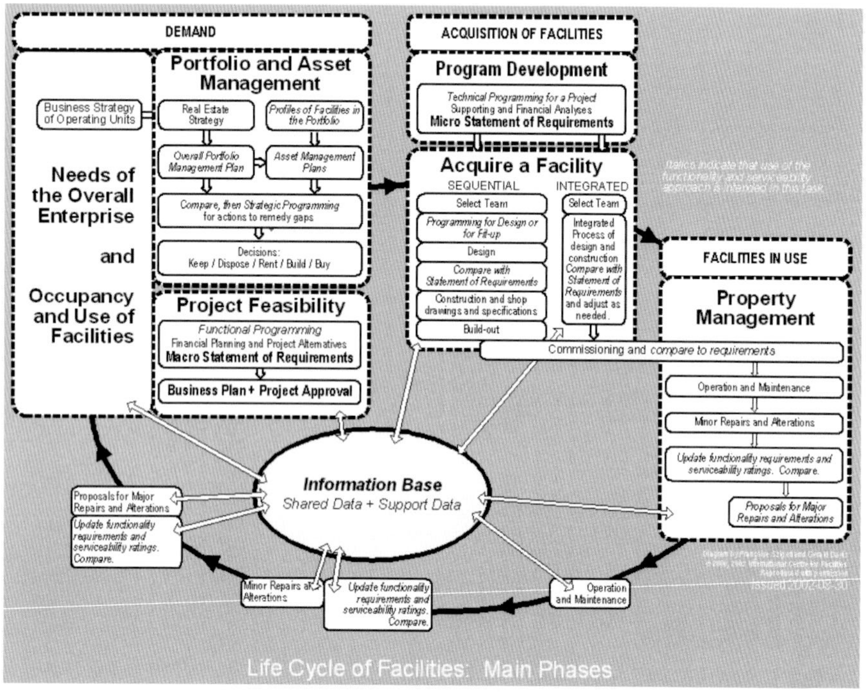

6.5
Industry Foundation Classes — life-cycle workflow structure.[19]

software tools used by industry participants[17,18] (figure 6.5). Although the latest version of the IFC (IFC2x 2nd Edition) captures significant data information of the complete building model, the goal of promoters and developers to have specifications for every class of building element, required to support data-sharing between all software tools used by building professionals, poses a challenge as it is an extensive task to build such a comprehensive data model.

To achieve interoperability among simulation tools during the building design and construction stages, the process-driven integration approach suggests there must be more than just data exchanges supported by a common building representation. Performance simulation needs a framework which allows it to be called upon at the right time and for the right design decision. The availability of domain-specific tools by themselves requires additional functionalities and frameworks in order to play a better role in design evolution. This approach focuses on the effective use of performance simulation tools in the design process, with full participation from the design team including the expert consultants. In 2001, a team of researchers from Georgia Institute of Technology, Carnegie Mellon University, and the University of Pennsylvania began work on the Design Analysis Integration Initiative (DAI) project in an attempt to develop credible solutions for the integration of building performance analysis tools in the building design process.[20] It intended to capitalize on the efforts already invested in the development of building product models — the IAI–IFC effort described above — without making limiting assumptions about the design process or the logic of the design analysis interaction flow. This process-driven integration approach is still in its early stages and will require more research and development to illustrate its full potential to solve the problem of integration.

PERFORMANCE TECHNIQUES AND METHODS

To increase the utility of the simulation in design, several methods and techniques are being incorporated. Decision support environments and optimization in the models, as well as user-friendly interfaces combined with visualization techniques, are the main areas of development in regard to performance-based simulation for architectural design.

Research in optimization and decision support environments began two decades ago. Computational algorithms were used to develop systems that assisted designers in their activities by providing either guidance through advice or optimization using emerging Artificial Intelligence techniques. Knowledge-based systems and complex problem-solving methods were researched for their potential use as an aid for both expert and non-experts to perform and interpret simulations.[21-24] Their contribution has been significant in initiating the use of computational techniques to solve performance-based decision-making problems. In addition, this research has inspired renewed efforts in the areas of optimization and decision support environments. This work is dominated by the idea that design is a goal-oriented decision-making model, where goals are defined by desired performance values. The work benefited from rapid improvements in the field of optimization to solve complex problems in areas of numerical methods, solution strategies and the development of new algorithms.

Design optimization using performance simulation has been researched recently to test the applicability of specific algorithms in regard to their effectiveness. During the past few years, stochastic methods such as simulated annealing and genetic algorithms (GAs) have become prevalent. They have been applied to a wide range of problems, including thermal and lighting performance in buildings.[25,26]

Recently, investigations of gradient-based and derivative-free methods were also conducted. Gradient-based methods are efficient and their results are reliable except in cases of complex simulation.[27,28] On the other hand, derivative-free deterministic methods perform well for problems that suffer from simulation noise and are mostly suitable for small problems.[29] Because simulation-based optimization can be time consuming, approximate-based methods which use functions derived from simulation responses to partially guide the search during the optimization process have been recently utilized.[30-33]

Strategies that combine two or more methods have also been used to overcome problems associated with one particular method. An example of such an approach is combining genetic algorithms and pattern searches to derive a hybrid method to reduce computational run time in problems involving expensive simulations.[34]

Although optimization algorithms have been used in a variety of ways in building simulation, their full potential in design synthesis has not been fulfilled. The work conducted over the past few years demonstrates that design optimization can be used to improve building performance and provides rigor in the way the simulation tools can be utilized. An example of such an approach is presented in the following section.

DESIGN ANALYSIS VERSUS PERFORMANCE-BASED SIMULATION

Fundamental to simulation tool development are questions related to the user and the problem the tools need to solve. The answers will orient the development to take a certain direction. Currently, simulation tools are analysis-based and support a variety of design activities. Rethinking the use of these tools from analysis to performance-based active design support has not been explored fully. In addition, the full potential of the recent developments in problem-solving techniques and visualization has not been realized. Incorporating such issues has the potential to allow active design support to exist. Taking advantage of these developments, as well as partially automating some of the simulation processes, will afford simulation tools a larger role in design activities.

An illustrative example of such research is a project we recently conducted that integrated computational fluid dynamics and GAs.[35] The project uses thermal and ventilation performance criteria as a means to generate creative design alternatives. This process requires an in-depth knowledge of the problem and considerable expertise. Each design change made to space geometry or its boundaries requires the user to re-model, re-mesh and subsequently re-compute the airflow. The trade-off between different design changes generally remains obscured because of the complexity of the model. As a result, the design space is difficult to explore systematically and solutions can be overlooked. Evolutionary techniques coupled with visualization were used as a solution to such a problem.

Evolutionary algorithms have traditionally been used to solve optimization problems. In addition, they can be used as a design aid. The evolutionary approach is a "generate-and-test" approach that corresponds well to the procedures for design synthesis and evaluation in the design process. In an evolutionary-based generative process, design representations are specified as a set of parameters and as a corresponding set of constraints. Generative design describes a broad class of design where the design instances are created automatically from a

6.6
Performance-based Design Evolution software interface.

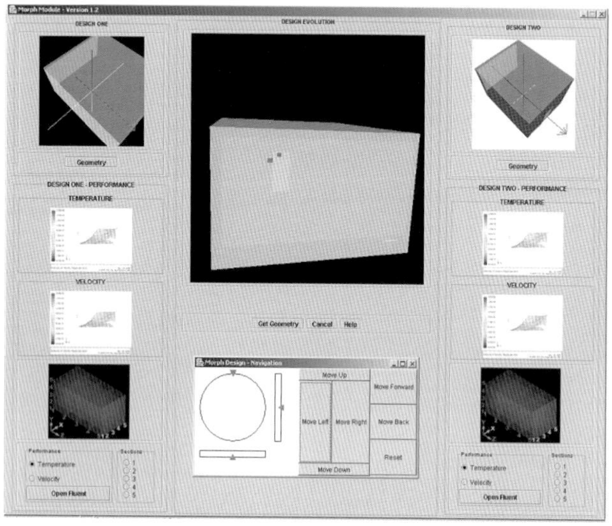

high-level specification. Design evolutions can be used as an aid in stimulating the designer creativity. The advantage of such an evolutionary approach is the creation of diverse sections of the state space that meet performance targets and increase the possibility for discovering a variety of potential solutions by providing a larger search space for designers to interact.

The project demonstrated the potential of exploring and visualizing the design evolution and its form generation based on a set of performance targets. It demonstrated how a performance-based design evolution model allows efficient exploration of design alternatives using a four-layer approach: design evolution, performance evaluation, morph visualization and design evaluation. The design evolution uses a GA module that generates the shape of the design instances. To explore the design space using GAs, the design variables are represented as an array of allele sets or "space genome". For each design change, a CFD analysis is automatically performed to evaluate the thermal and ventilation efficiency performance, and the results are fed back into the GA module. This process runs recursively until the best designs are returned. To capture the system's solutions and in-between instances as the solution is evolving, the system is integrated with a morphing module. This module allows the diverse instances of designs to be visualized along with their performance in order to enhance the user's potential for design discovery and to provide aid within the design process. During morphing, users can intervene to stop the process and select an instance based on its form (figure 6.6). This design instance can then be evaluated for its performance using an integrated CFD engine. By integrating an optimization module that creates discrete instances of design with a morphing module, the system was able to provide an example of continuous evolution of optimization. Although reversing the cycle from the design-analysis paradigm to a performance-based simulation-decision-support paradigm has been previously researched, the full potential of this shift has not been realized. Despite the advances in computational techniques that can be used in the development of such tools, this area is still in its infancy. This is partly due to the multidisciplinary expertise required for such development and the fragmentation of the building industry, which makes it difficult to establish a unified voice to articulate such requests.

DATA VISUALIZATION AND INTERFACE DESIGN

Most of the available tools provide a one- or two-dimensional representation of the data derived from a building performance simulation. This has always been an important challenge, as only experts can precisely understand the data and hence are always required to interpret them. Consequently, this introduces the problems of time and cost, not only in terms of hiring these experts but also in establishing communication among the participants. This communication is not only dependent on their physical presence, it also involves issues of representation as well as of semantics. Advances in visualization led to new developments in simulation. Different technologies have made it possible to create environments that are virtual or augmented. Although immersive building simulation is still in its research and development stage, virtual and augmented environments have been used in a variety of areas in

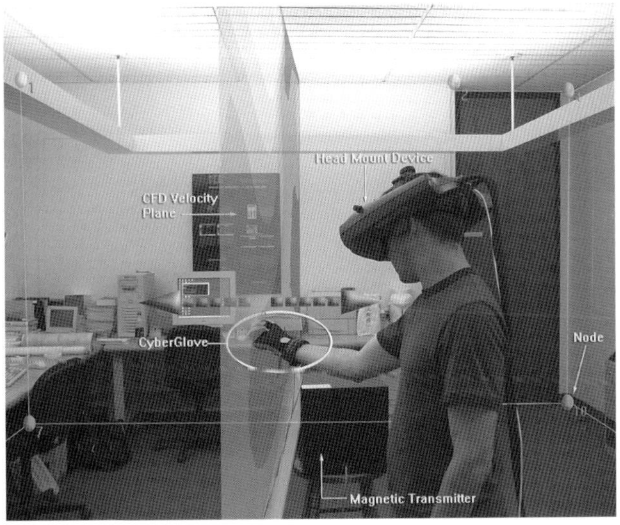

6.7
Immersive building simulation.

relation to buildings. This includes the extension of visual perception by enabling the user to see through or into objects,[36] such as maintenance support for visualizing electrical wires in a wall or construction grids.[37] Other applications include structural system visualization,[38] augmented outdoor visualization[39] and a collaborative design process.[40]

In the area of *immersive building simulation*,[41] only a few projects have been developed — some of which are related to the post-processing of CFD data,[42,43] augmented simulations,[44] building and data representation,[45] building performance visualization,[46,47] immersive visualization for structural analysis,[48] and interactive immersive building simulation[49,50] (figure 6.7).

The few studies conducted in this area illustrate the potential application of such research. Despite the prevalence of challenges related to software, hardware and the knowledge required to apply this to the building design, immersive environments provide opportunities which are not available using current simulation models and interactions, and it extends on the three-dimensional performance information visualization of buildings. It presents a new way of interfacing with the built environment and controlling its behavior in real time. This will become increasingly evident as additional techniques (e.g. optimization) lend more power to these environments as users navigate through them and interact with their elements.

CONCLUSIONS

The use of performance simulation in architectural design is on the rise. This is due mainly to the increase of computing power and the maturing of the building simulation field. The full integration of simulation within the design process is far from complete. However, the building industry, including architects, is aware of the need for the better integration of these tools into the life cycle of the building. Many of the developments presented in this chapter are a testimony to the rapid use of these tools, methods, and technology into design. However, only elite practices are currently taking advantage of the recent developments in performance simulation.

The tools are predominately of analysis and are not for analysis and synthesis. Challenges related to performance simulation accuracy, which is influenced by factors such as user interpretations and interventions to variations in simulation variables and behavioral uncertainties and validations, are still being investigated. Although a collective effort to develop standards for integration is underway, the chapter illustrates that this is a non-trivial issue. As the demand to advance performance simulation increases, new paradigms have begun to emerge and will influence building design and construction. These paradigms are indications of the field's various directions, some of which are purely demand-oriented, while others forecast the needs of the field in an attempt to enhance it. This chapter has illustrated examples of such paradigms which include issues of process-driven interoperability, the influence of optimization techniques into the design of new tools, and new means of interaction between the users and simulations. Although such work is underway, we are far from utilizing the full potential of the advancement in the computational fields that can impact the way architecture is practiced and experienced.

NOTES

1 A. M. Malkawi, "Developments in Environmental Performance Simulation" in *Automation in Construction Journal* 13:4, 2004, pp. 437–445.
2 DOE online directory of energy related tools for buildings can be accessed at: http://www.eren.doe.gov/buildings/tools_directory/
3 D. B. Crawley, L. K. Lawrie, C. O. Pedersen, F. C. Winkelmann, M. J. Witte, R. K. Strand, R. J. Liesen, W. F. Buhl, Huang, R. H. Henninger, J. Glazer, D. E. Fisher and D. Shirey, "EnergyPlus: New, Capable and Linked" in *Journal of Architectural and Planning Research* 21:4, 2004.
4 G. C. da Graca, Q. Chen, L. R. Glicksman and L. K. Norford, "Simulation and Wind-driven Ventilative Cooling Systems for an Apartment Building in Beijing and Shanghai" in *Energy and Buildings* 34: 1, 2002, pp. 1–11.
5 H. Huang and F. Haghighat. "Modeling of Volatile Organic Compounds Emission from Dry Building Materials" in *Building and Environment* 37: 12, 2002, pp. 1349–1360.
6 S. M. Lo, K. K. Yuen, W. Z. Lu and D. H. Chen. "A CFD Study of Buoyancy Effects on Smoke Spread in a Refuge Floor of a High-rise Building" in *Journal of Fire Sciences* 20: 6, 2002, pp. 439–463.
7 G. Forney, D. Madrzkowski and K. B. McGrattan. "Understanding Fire and Smoke Flow Through Modeling and Visualization" in *IEEE Computer Graphics and Applications,* Los Alamitos, CA: IEEE Publications, 2003.
8 Fluent Inc., *Fluent User's Guide,* New Haven, USA, 2003.
9 Y. Jiang and Q. Chen. "Effect of Fluctuating Wind Direction on Cross Natural Ventilation in Building from Large Eddy Simulation" in *Building and Environment* 37: 4, 2002, pp. 379–386.
10 Q. Chen and Z. Zhai, "The use of CFD Tools for Indoor Environmental Design" in A. M. Malkawi and G. Augenbroe (eds), *Advanced Building Simulation,* London: Taylor & Francis, 2004.
11 Forney *et al.*, op.cit.
12 Malkawi, op.cit., "Developments in Environmental Performance Simulation."
13 K. Papamichael, J. La Porta and H. Chauvet, "Building Design Advisor: Automated Integration of Multiple Simulation Tools" in *Automation in Construction Journal* 6: 4, 1997, pp. 341–352.
14 N. H. Wong and A. Mahdavi, "Automated Generation of Nodal Representations for Complex Building Geometries in the SEMPER Environment" in *Automation in Construction Journal* 10: 1, 2000, pp. 141–153.
15 Malkawi, op.cit., "Developments in Environmental Performance Simulation."
16 ISO-STEP, ISO 10303 — the industrial automation systems and integration standards can be accessed at: http://www.iso.ch
17 The IAI can be accessed at: http://www.iai-international.org/iai_international
18 V. Bazjanac and D. B. Crawley, "Industry Foundation Classes and Interoperable Commercial Software in Support of Design of Energy-efficient Buildings" in *Proceedings of Building Simulation,* IBPSA, 1999, pp. 661–668.
19 IAI, op.cit.
20 G. Augenbroe, P. de Wilde, H. J. Moon, A. Malkawi, R. Brahme and R. Choudhary, "The Design Analysis Integration (DAI) Initiative" in *Proceedings of Building Simulation,* Eindhoven, The Netherlands: IBPSA Publications, 2003.
21 A. D. Radford and J. S. Gero, *Design by Optimization in Architecture and Building,* New York: Van Nostrand Reinhold, 1988.
22 J. Pohl, L. Myers and A. Chapman, "A Knowledge-based System for the Design of Passive Solar Buildings" in *ASHRAE Transactions* 96: 2, 1990.
23 E. Shaviv and Y. E. Kalay, "Combined Procedural and Heuristic Method to Energy-conscious Building Design and Evaluation" in Y. E. Kalay, *Evaluating and Predicting Design Performance,* New York: Wiley, 1992.
24 A.M. Malkawi, *Building Energy Design and Optimization: Intelligent Computer-aided Thermal Design,* PhD dissertation, Georgia Institute of Technology, 1994.
25 J. A. Wright, H. A. Loosemore and R. Farmani, "Optimization of Building Thermal Design and Control by Multi-criterion Genetic Algorithm" in *Energy and Buildings* 34: 9, 2002, pp. 959–972.
26 R, Choudhary and A. M. Malkawi, "Integration of CFD and Genetic Algorithms" in *Proceedings of the Eighth International Conference on Air Distribution in Rooms,* Technical University of Denmark and Danvak, Denmark, 2002.
27 H. Jedrzejuk and W. Marks, "Optimization of Shape and Functional Structure of Buildings as well as Heat Source Utilization Example" in *Building and Environment* 37, 2002, pp. 1249–1253.
28 R. Choudhary, A. M. Malkawi and P. Y. Palambros, "A Hierarchical Design Optimization Framework for Building Performance Analysis" in *Proceedings of Building Simulation,* Eindhoven, The Netherlands: IBPSA Publications, 2003.
29 M. Whetter and E. Polak, "A Convergent Optimization Method using Pattern Search Algorithms with Adaptive Precision Simulation" in *Proceedings of Building Simulation,* Eindhoven, The Netherlands: IBPSA Publications, 2003.

30 Choudhary *et al.*, op. cit., "A Hierarchical Design Optimization Framework for Building Performance Analysis."
31 M. Whetter and E. Polak, op.cit.
32 A. M. Malkawi, R. S. Srinivasan, Y. K. Yi, and R. Choudhary, "Performance-based Design Evolution: The Use of Genetic Algorithms and CFD" in *Proceedings of Building Simulation*, Eindhoven, The Netherlands: IBPSA Publications, 2003.
33 R. Choudhary, "A Hierarchical Optimization Framework for Simulation-based Architectural Design," PhD dissertation, University of Michigan, 2004.
34 M. Whetter and J. Wright, "Comparison of a Generalized Pattern Search and a Genetic Algorithm Optimization Method" in *Proceedings of Building Simulation*, IBPSA, 2003.
35 Malkawi *et al.*, op.cit., "Performance-based Design Evolution: The Use of Genetic Algorithms and CFD."
36 G. Klinker, D. Stricker, D. Reiners, R. Koch and L. Van-Gool, "The Use of Reality Models in Augmented Reality Applications. 3D Structures for Multiple Images of Large Scale Environments" in *Proceedings of the European Workshop SMILE*, 1998, pp. 275–289.
37 A. Retik, G. Mair, R. Fryer and D. McGregor, "Integrating Virtual Reality and Telepresence to Remotely Monitor Construction Sites, a ViRTUE Project" in *Artificial Intelligence in Structural Engineering*, 1998, p. 459.
38 S. Fiener, A. Webster, T. Kruger, B. MacIntyre and E. Keller, "Architectural Anatomy" in *Presence*, no. 4, 1995, pp. 318–325.
39 M. O. Berger, B. Wrobel-Dautcourt, S. Petitjean and G. Simon, "Mixing Synthetic and Video Images of an Outdoor Urban Environment" in *Machine Vision and Applications*, no. 11, 1999, pp. 145–159
40 P. Frost and P. Warren, "Virtual Reality used in a Collaborative Architectural Design Process" in *Proceedings of Information Visualization, IEEE International Conference*, 2000, pp. 568–573.
41 A. M. Malkawi, "Immersive Building Simulation" in A. M. Malkawi and G. Augenbroe (eds), *Advanced Building Simulation*, London: Taylor and Francis, 2004.
42 V. J. Shahnawaz, J. M. Vance, and S. V. Kutti, "Visualization of Post-processed CFD Data in a Virtual Environment" in *Proceedings of DETC99*, 1999, pp. 1–7.
43 A. M. Malkawi and E. Primikiri, "Visualizing Building Performance in a Multi-user Virtual Environment" in *Proceedings of the ACSA International Conference*, 2002.
44 A. M. Malkawi and R. Choudhary, "Visualizing the Sensed Environment in the Real World" in *Journal of Human-Environment Systems*, no. 3, 1999, pp. 61–69.
45 M. Pilgrim, D. Bouchlaghem, D. Loveday and M. Holmes, "A Mixed Reality System for Building Form and Data Representation" in *IEEE*, 2001, pp. 369–375.
46 A. M. Malkawi and R. Choudhary, op.cit., "Visualizing the Sensed in the Real World."
47 Pilgrim *et al.*, op.cit.
48 N. Rangaraju and M. Terk, "Framework for Immersive Visualization of Building Analysis Data" in *Information Visualization IEEE International Conference*, Los Alamitos, CA: IEEE Publications, 2001, pp. 37–42.
49 A. M. Malkawi, R. S. Srinivasan, B. Jackson, Y. K. Yi, K. H. Chan and S. Angelov, "Interactive, Immersive Visualization for Indoor Environments: Use of Augmented Reality, Human-Computer Interaction and Building Simulation," *Information Visualization IEEE International Conference*, 2004.
50 A. M. Malkawi and R. S. Srinivasan, "Multimodal Human-Computer Interaction for Immersive Visualization: Integrating Speech-gesture Recognitions and Augmented Reality for Indoor Environments," *IASTED Computer Graphics and Imaging (CGIM) Conference*, IASTED, Canada, 2004 (in press).

7

A FRAMEWORK FOR RATIONAL BUILDING PERFORMANCE DIALOGUES

GODFRIED AUGENBROE

7

A FRAMEWORK FOR RATIONAL BUILDING PERFORMANCE DIALOGUES

GODFRIED AUGENBROE

This chapter deals with an engineering perspective on the development and maintenance of buildings. Performance plays a major role in the expectations expressed by owners and occupants, and the fulfillment of them by designers and building operators. The "disconnects" between expectations and fulfillment are rampant throughout the building delivery process, and the better matching of the two is considered an important target of the building industry, in order to become more client-driven and to provide better value overall thereby guaranteeing customer satisfaction.[1]

The need to move the industry in this direction has fostered the introduction of a performance-based building method.[2] These methods focus on the generation of performance-based statements of requirements and on the management of a transparent process that guarantees their fulfillment. Essentially, this requires better methods and tools to support the communication between designers, engineers and building managers. An essential part of this communication is the "dialogue" that explicitly communicates expectations and fulfillments between a demand and supply party. Two dialogues are of particular importance: (1) the architecture and engineering (A/E) procurement dialogue, which deals with the way services are procured by the design team to engineer building systems that meet functional needs and client expectations; and (2) the tenant/facility manager dialogue, which deals with the proper maintenance and management of the facility in a way that meets the expectations of the occupant, owner or portfolio manager and that provides and maintains maximum value from the facility for all stakeholders.

Little elaboration is needed on the fact that the current ways and methods to maintain and support these dialogues are hampered by a range of deficiencies which need to be overcome. Among these deficiencies we highlight the "asymmetric ignorance" between the communicating partners, the lack of transparent requester-provider roles in the process,[3] the lack of objectively quantifiable expressions of requirements, and a lack of the proper assessment tools to ascertain whether expectations have been fulfilled by a proposed design (in dialogue 1) or by a proposed tenant/building allocation (in dialogue 2).

In addition to the failure of the community at large to add quantified elements to the dialogues, there is also an apparent cultural resistance, mainly from the side of the architectural designer, where many seem to believe that architectural performance cannot be measured. The habitual conjecture in this discussion is that many aspects of performance can only be interpreted based on qualitative judgments. It is also debated whether these judgments have to rely on unpredictable manifestations of the design in its not fully determinable context of use. Unfortunately, these judgments are biased by the "value system" of the one who measures them and, at best, lead to some measure of quality, i.e. "a set of characteristics that are perceived to contribute to value."[4] In the following sections it is argued that this is not good enough and that many aspects of buildings can and should be measured objectively. The performance characteristics that are most amenable to an objective statement are those that relate to the functions that the building, or one of its (sub)systems, are designed to perform. Instead of subjective quality, objective "utility" should be introduced to represent an aggregation of objectively measurable performance characteristics. The aggregation is performed over systems and functions.

Traditionally, the dialogues mentioned above have been cast in prescriptive terms, i.e. specifying aspects of the solution rather than the expected performance of the solution. Building codes and regulations have long contributed to this by basing their approach on prescriptive specification methods. However, this is now no longer the case, as many countries are moving parts of their regulations and standards to the performance domain.[5]

Different methods have come into existence to support the dialogue between the demand and supply side. An interesting method which is in use today is documented in the set of ASTM (American Society for Testing and Materials) standards for Whole Building Functionality and Serviceability.[6] The standards lay out a methodology known as "Serviceability Tools and Methods" (ST&M) which can be used through various stages of the project delivery process. ST&M provides a method to measure how well a design proposal, or an existing facility, meets the requirements of the stakeholders, for about 100 different performance topics. The current ASTM standards provide a broad-brush, macro-level method, appropriate for strategic, overall decision-making. They deal both with demand (occupant requirements) and supply (serviceability of buildings).[7] The expression of demand and supply is captured per topic in so-called scales, each ranging from the lowest score (0) to the highest score (9). The scores reflect a rating of the (design of a) facility, both from the demand as well as the supply perspective. What is rated on the supply side is the "fitness" of the facility to meet a given need, hence the term serviceability. More information can be found in the work done by Szigeti and Davis.[8]

Whereas the ST&M method provides a coarse method geared towards the corporate use (many of the topics relate to the serviceability of organizational needs), and is based on a scoring by trained users, we are focusing on the introduction of a fine-grained set of performance quantifiers that rely solely on metrics derived from first order physical principles, i.e. ruling out the interpretation of experts. We argue that in doing so, an expanding set of performance metrics is added as quantified elements in the expectation-fulfillment dialogue. One must realize that quantifications are only useful when they are very precisely defined; otherwise they lead to misinterpretation and do not support a rational dialogue.

This chapter introduces a framework for the rational expressions of building performance, based on the notion of objectively quantifiable measures. It will discuss how these measures can be viewed as a set of uniquely defined "performance indicators," whose quantifications are linked to observable building behavior in specific (virtual) experiments. The quantification perspective in this chapter is predominantly engineering driven. A system-theoretical basis is explored to express both requirements and assessed performances at varying levels of building systems granularity, and its potential to support design dialogues is verified.

The treatment is positioned to support rational decision-making during the different stages of building delivery and use. The focus is specifically on the fulfillment of client expectations during design evolution and on the assessment of building tenant combinations over the service life of the building. A toolkit for the rapid quantification of measures for new and existing buildings is introduced.

The following sections will discuss the methods to define performance metrics that lead to objective and rational elements in the dialogues addressed above. The first step towards this is a framework that enables the identification of systems and their contribution to desired functions, and this is the subject of the next section.

A BUILDING SYSTEMS CLASSIFICATION FRAMEWORK

The rationalization of building performance in dialogues between stakeholders requires the systematic discovery of performance domains and their specialization to the level where individual measures of performance can be addressed. Such individual measures of performance will be termed performance indicators (PIs).

There is no established theory for the discovery and definition of PIs. Wim Gielingh in his General AEC Reference Model[9] hypothesizes a design process in which the design problem is reduced into smaller manageable functional units (FUs). For each of these FUs, a designer looks for a technical solution (TS) that satisfies the functional requirements of the FU. When no solution is found, the design problem is restated with a new set of functional requirements, or a new solution is developed to meet the functional requirements. Solutions to more complex design problems are obtained by reducing the design problem into smaller design problems. These smaller design problems are themselves new FUs that require new TSs. This iterative process continues until solutions have been found for all FUs. PIs can be thought of as the "meat" between the FUs and TSs in Gielingh's model (figure 7.1). Although Gielingh's work was ground-breaking in its conceptualization, it was too "idealized" and has, as a result, stayed far removed from practical applications.

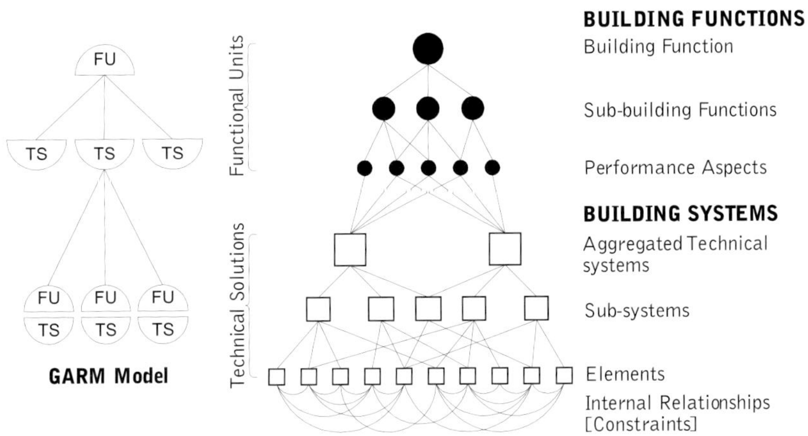

7.1
Overview of a functional systems approach.

Another systems approach[10] to building design divides the activity into two distinct but related views: (1) the functional view, which addresses the functional aspects of the building design; and (2) the building systems view, which deals with systems that are designed or selected to fulfill the functional requirements of the building. The main function of a building can be decomposed into lower-level functions, such as safety, habitability and sustainability. Subsets of these lower-level building sub-functions have been identified by various standards bodies, such as the International Organization for Standardization (ISO). The lower half of figure 7.1 shows the aggregation of technical systems from subsystems, until we reach the level of the well know major "functional systems," such as lighting, fire safety, etc. These systems contribute to one or more of the required building functions. Elements are the smallest decomposition of a building system and may belong to one or more subsystems. There exists a constraint relationship within elements of a system and between elements of different subsystems. This systems view is illustrated in figure 7.1.

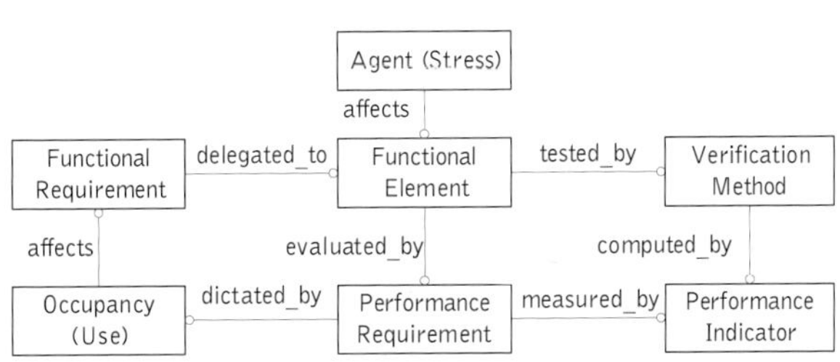

7.2
The relationship among concepts in the performance approach.[14]

The goal of the performance approach to building design and maintenance is to provide a mechanism by which one can measure how well a building system is capable of fulfilling the functional requirements of the building that it serves. In order to do this, the performance approach to building design addresses building requirements in terms of performance rather than properties. In other words, based on test results rather than prescribed dimensions, specific attributes or specific materials. This allows any material or product that meets the performance requirements to be considered for use.[11] There are seven main concepts in the performance approach (figure 7.2).[12,13]

1. Goal (occupancy) is usually a qualitative statement that addresses the needs of the user-consumer and determines the required level of performance.
2. Functional requirements are the mandatory requirements that must be fulfilled to ensure users are satisfied with the facility.
3. Performance requirements are the user requirements expressed in terms of the performance of a product. Performance requirements are measured by PIs.
4. PIs are quantifiable indicators that adequately represent a particular performance requirement. A PI, by definition, is an agreed-upon indicator that can be quantified using a verification method.
5. Verification methods are used to evaluate whether the performance requirement has been met. Verification methods can be experiments (tests), calculations, or a combination of both. Each PI has its own verification method. A number of verification methods may exist for measuring the same performance requirement.
6. Functional elements are parts of a building that fulfill one or more functions to meet the user requirements. Functional elements can be materials, construction, products from a product catalogue, and even non-physical entities such as building spaces.
7. Agents alter the ability of functional elements to satisfy the functional requirements.

The first two concepts provide the functional goals that must be achieved. The relationship between the performance aspects and the building systems defines the level of performance that is required to achieve that goal. PIs are used to quantify the performance requirement in order to determine whether a particular design satisfies the performance requirement. The context information required by the verification method is provided by the functional system that represents all the properties and all the relationships of all the elements in that system.

Although these existing frameworks provide a useful model for the top-down discovery of PIs, they are not specific enough to be deployed in the general building case. One of the main reasons for this is that no standard decomposition of functional requirements and no set of ready-made subsystems exist in the bottom half of figure 7.1. With the infinite variability that characterizes the current style of building design management, the framework does not offer a useful way out. A more bottom-up oriented approach to the discovery of PIs can provide a better way to achieve more precision in the mentioned dialogue. Designers have opportunities to use the performance approach at all stages of the design process. Not only does the performance approach enable the designer to discipline his own contribution but, perhaps more importantly, it provides the potential for a clearer definition of responsibilities when the analysis part of the process is delegated.

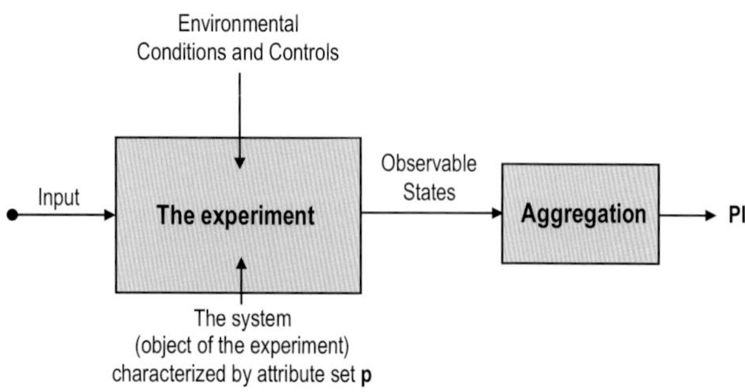

7.3
The components of the AF description.

Such a bottom-up approach is by nature domain-specific, dealing with a specific analysis expertise and a limited set of functional requirements and a subset of the systems that contribute to that functionality. In the next sections, we will focus on functional requirements concerning energy usage, thermal comfort and daylight usage. This requires focusing on those building systems that contribute to these functions, such as the heating and cooling systems and the building enclosure system. As figure 7.1 demonstrates, these building subsystems can be viewed as compositions of standard components which are not going to be redesigned in every building. It is at the connections of systems and functions that PIs are introduced as a way to express their contribution to reaching a desired functionality. It is important to point out that there is no unique way to measure the contribution of a particular system to an overall function. Many different ways to measure performance (i.e. in different experiments) can be introduced, each with its own merit. So, each connection in figure 7.1 may have multiple PIs attached to it.

A practical bottom-up discovery of PIs may be attempted by closely investigating the type of stakeholder interactions with respect to expected and fulfilled performances that take place in real projects. Different PIs can be found in different types of performance evaluation requests.

Although the above approach will reveal the primary connections between functions and systems, and thus point towards the need for one or more PIs that "objectify" this relationship, we will need a conceptual framework to define the PIs themselves. This is the subject of the next section.

PERFORMANCE INDICATORS

This section introduces the concept of Analysis Functions (AFs) as a formal quantification method of a PI.

Figure 7.3 shows the basic notion of an AF as a mapping of experimental input variables, environmental and control variables and system properties (**p**) to a quantified PI through a specific aggregation procedure and illustrates all the components that are part of a full AF model. As an AF model uniquely describes the input and usage variables, the PI is dependent on the system properties, i.e. PI(**p**).

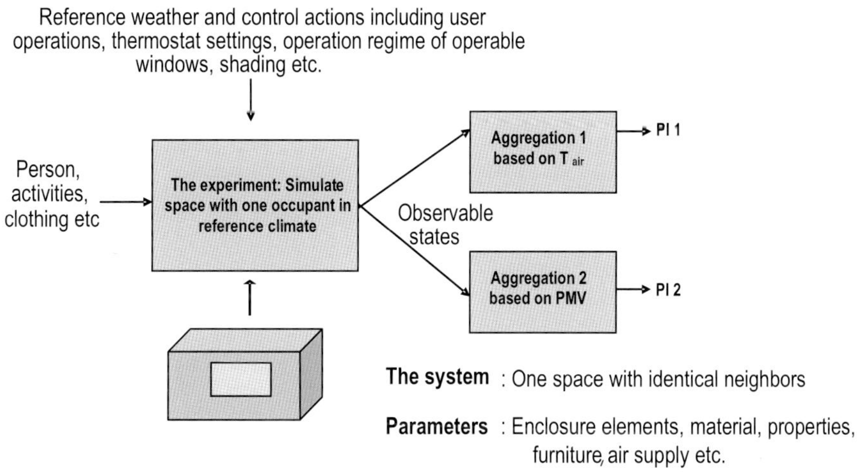

7.4
AF realizations for thermal comfort PI.

Many types of performance (e.g. thermal comfort) can be measured in different ways, and can be used interchangeably by the expert community, e.g. (i) the number of hours per year that the room air temperature exceeds a certain threshold, or (ii) the yearly Predicted Mean Vote (PMV) distribution.[15] Both methods of measuring thermal comfort performance intend to measure the same performance aspect, and set up the same virtual experiment but use different metrics (a different aggregation method leads to different metrics). This results in the definition of two distinctly different PIs. However, in this case, the two PIs would require the same virtual experiment (in all likelihood performed by the same software) but would use different ways of aggregating the outputs to quantify the two different measures.

Elaborating on the thermal comfort performance example, one can say that thermal comfort performance is delivered by the "comfort control system," composed of the heating, cooling, control (thermostats and control system), and enclosure systems. Different quantifications differ in the way that temporal temperature information is translated into a measure, i.e. in the choice of aggregation method. The different PIs are based on the same type of experiment: a human is placed in a certain location in a given space of the building in the local climate. The experiment variables determine thermostat control, ventilation actions (opening of windows) and observer properties, such as the activity level and clothing, as suggested in figure 7.4.

It should be noted that the AF must be described formally in order to rule out any judgment or interpretation issues during the quantification. This formal description will be referred to as the "AF model," constituting a full description of the experiment. The AF model for a thermal comfort PI formally describes all the entities that are needed to calculate the relevant state variables that determine thermal comfort. In the case of PMV, there needs to be an internal air zone that has an average air temperature, and there needs to be surfaces that have temperatures that can be used to calculate a mean radiant temperature. Also, occupants need to be defined that have a metabolic rate and clothing value. However, if the decision had been made to base the thermal comfort analysis function on a different measure, for instance the use of degree hours for the air temperature, then there would not have been a need to include any occupant or occupant properties, and the treatment of the surfaces might have been different. Note that different AFs for thermal comfort, like a PMV-based and a degree hour-based function, can coexist in the dialogue, and they should both be defined independently of their realization.

Going back to the different types of experiments that may be deployed, a couple of observations can be made. First, the mapping from (p) to behavior of the system is the key part of the quantification of the PI. It should be clear that the theoretical foundation of this mapping determines the reliability of a PI. In the case of a real experiment, the experimental set-up must be such that all disturbances are kept to a minimum and the monitoring noise does not significantly influence the measured behavior (as expressed through gathered state information). In a virtual experiment, one must be able to guarantee that the (simulation) tool's representation is adequate in order to accurately predict the behavior of the system. In thought experiments, biases must be avoided by clearly stated procedures. This can be accomplished by developing procedures that map the experimental variables onto a rating scale. Examples of these normative procedures are the already introduced serviceability rating methods developed by Davis and Szigetti[16] and the sustainability rating method LEED.[17] All the experiments provide PI(p) mapping based on a set of rules that describe the experiment in precise terms in order to perform the following two steps without any ambiguity:

> Step 1: Perform experiment which results in *Object_state (experiment_variables, object (p); t).*
>
> Step 2: Perform time and space aggregation over *Object_State*, resulting in *PI (experiment_variables, p).*

Based on a formal description of the measure (the above two steps) in an AF model, it will become possible to automate the execution of the AF. For this it is necessary to translate the AF model into a formal computer readable representation. This representation should capture the object (geometry, material properties, etc.), the experimental conditions and control settings, as well as the aggregation procedures. A data modeling language can adequately capture this information. Recent work that we conducted[18] shows an example of a workbench to accomplish this, using EXPRESS-G as a graphical language which allows entities and their relationships to be presented. The diagrams can be translated to an EXPRESS model, a computer interpretable version of EXPRESS-G, used in the implementation stages of the computer systems. The conceptual models structure all the necessary and sufficient information of the experiment.

The selection of which PIs should be embedded in a performance vocabulary is ultimately a choice of elegance and efficiency, and should be driven by the sound application of the 80-20 rule, which suggests that one aims at a relatively small set of PIs that requires moderate effort to develop and which still covers most of the (routine) dialogues in A/E procurement. We assume that a manageable set of concise PIs can be designed with that purpose in mind.

As this section suggests, the derivability of a necessary and sufficient set of PIs is as yet unproven in practice. Hence there is no guarantee that a limited and manageable set can be derived to express the majority of PIs needed in rational decision-making. A more practical approach is to develop toolkits that quantify a dedicated small set of PIs that have particular significance to a design office, a client, a tenant organization or corporate owner. The next section shows an example of this type of development for a tenant organization with emphasis on portfolio management.

APPLICATION: A PERFORMANCE TOOLKIT FOR PORTFOLIO MANAGEMENT

The need for continuous monitoring of the technical performance of buildings over their lifetime becomes obvious if one realizes that buildings undergo drastic changes and redefinition of their internal client processes. These changes are caused by internal reorganizations, refurbishments, change of tenants, etc. The natural degradation of the technical systems is another reason to track the performance of the assets in the portfolio and to link it to maintenance scheduling.

Using the approach of the previous sections, a set of performance measures was developed that provide cost-effective, quantitative predictions and assessments of how well buildings enable specific client functions.

The first version of the toolkit provides metrics for performance tracking in the areas of energy, lighting, thermal comfort and maintenance. The metrics are embodied in a set of PIs based on standardized and normative calculation routines that assess a building quickly, reflecting its actual use. Based on the theories in biophysics and physiology, these measures quantify the performance of a building system in producing a desired condition, related to an activity or need of the tenant or any other stakeholder. The metrics allow the evaluator to understand how multiple systems interact to produce a given level of building performance. The resulting PIs can be used to formulate requirements as well as to quantify actual performance.

The set of PIs have been harnessed in an operational building performance assessment toolkit specifically adapted for the use by tenant organizations for portfolio management. The toolkit is, however, equally relevant to design firms in the A/E procurement stages and for

Table 7.1
Performance indicators (PIs).

Aspect	Function	PI	Meaning
Energy	Energy	PI 1–7	Heating, cooling, humidifying, lighting, pumps, fans, hot water in MJ
Lighting	Energy efficiency	PI 1	Electric lighting energy consumption over required illuminance level in kWh/m^2 •year •lux
		PI 2	Luminous efficacy of luminaires in LER (Lumens/watt)
		PI 3	Daylighting autonomy: percentage of hours without requiring an artificial lighting
	Visual comfort	PI 4	Ratio of task illuminance as installed and as required
		PI 5	Outward visibility (view to outside): percentage of occupants who can see the outside from their workplaces
		PI 6	Daylighting glare avoidance: percentage of office hours in discomfort range (Daylighting Glare Index •24, just uncomfortable)
		PI 7	Shading devices for glare avoidance (under development)
Thermal comfort	Air diffusion	PI 1	Percentage of occupants in comfort in ADPI (Air Diffusion Performance Index)
	Asymmetric thermal radiation due to hot/cold glazing	PI 2	Hourly average Predicted Percentage of Dissatisfied occupants (PPD) during office hours over a year
		PI 3	Percentage of hours where the PPD is in the comfort range (10%)
		PI 4	Average of PPD where the PPD is not in the comfort range
	Cold draft caused by glazing	PI 5	Hourly average PPD during office hours over a year
		PI 6	Percentage of hours where the PPD is in the comfort range (10%)
		PI 7	Average of PPD where the PPD is not in the comfort range
	Diversity	PI 8	Average PPD of workers in different activities and clothing levels
	Zoning	PI 9	Heating airflows variation in different rooms in a thermal zone
		PI 10	Cooling airflows variation in different rooms in a thermal zone
	System's capacity and response time	PI 11	The seconds required to increase the zone temperature by 1°C at the peak load time
Mainte-nance	Efficiency	PI 1	Building Performance Indicator (BPI), scaled from 0 to 100
		PI 2	Maintenance Efficiency Indicator (MEI)
	Business and organization	PI 3	Manpower Sources Diagram (MSD): a ratio of in-house and outsourcing expenditures
		PI 4	Managerial Span of Control (MSC): a ratio of a manager and subordinated personnel
		PI 5	Business availability in per cent: an available floor area over an entire floor area over year
		PI 6	Manpower Utilization Index (MUI) in per cent: a ratio of man-hours spent on maintenance and total available man-hours
		PI 7	Preventive Maintenance Ratio (PMR) in %: a ratio of man-hours spent on preventive maintenance and spent on total maintenance (preventive + corrective)
	Timeliness	PI 8	Urgent Repair Request Index (URI): occurrence/10,000 m^2
		PI 9	Failure frequency: occurrence/10,000 m^2 and unit repairing time in minutes
	Policy	PI 10	Maintenance productivity: state/$ (under development)

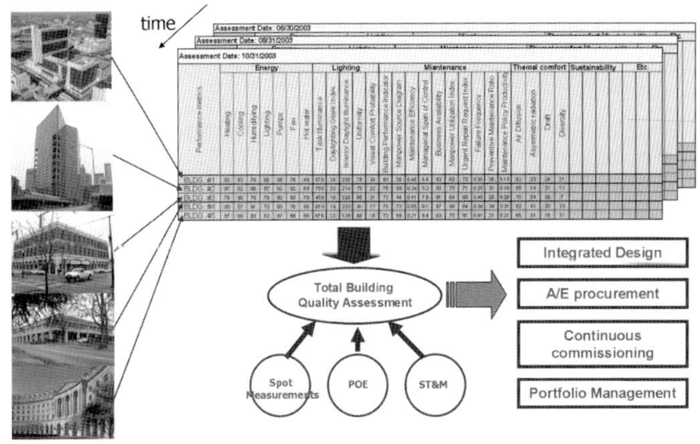

7.5
The landscape of building performance assessment in large organizations.

corporate owners who realize that buildings should be constantly monitored to guarantee a safe, healthy and productive environment for their occupants. For a government organization such as the General Services Administration (GSA), the use of the toolkit will deliver the data that underpin the day-to-day service management as part of GSA's tactical operations. Figure 7.5 shows how the buildings in the portfolio will be assessed on a regular basis (e.g. once a year). The gathered data will be entered and PI quantifications will be done automatically by the toolkit routines. The data will then be uploaded in a database where PIs are stored per category. At this point in time, the toolkit contains roughly 25 indicators. The uploaded data can be browsed and linked to other data about the monitored assets (pictures, floor plans, maintenance records, and others). The access to the records of the facility over previous years provides a way to inspect deterioration trends or to inspect sudden changes in performance, e.g. when a new tenant is allocated to the building. As should be clear from the above, PIs measure the building tenant combination, as the virtual experiment measures behavior relative to the functions and needs of an occupant. As figure 7.5 implies, all performance data are accessible for total building quality management. All other sources of data, such as local physical measurements, post-occupancy evaluation (POE) data, as well as design programs, ST&M records, etc., will be linked to enable data mining for the business intelligence gathering that is necessary to improve the design development, procurement and facility maintenance procedures.

Until now, the work has concentrated on four performance aspects: energy, lighting, thermal comfort, and maintenance. Some of the issues involved in generating the indicators and their benchmarking on real buildings are addressed below.

The list of Pis is shown in table 7.1.

To elucidate the development issues of the indicators, we will again take the example of thermal comfort. For its assessment, eleven Pis were developed to account for space air diffusion, asymmetric radiation, cold draft, occupant diversity, HVAC zoning and system's response (refer to table 7.1). It is important to note that all of these effects can be studied through full simulation of the office spaces. This, however, would defeat the purpose of a rapid repetitive evaluation. A full simulation would also introduce bias into the PI quantification, as no simulation can be performed without introducing assumptions and simplifications, often dependent on the type of simulation tool. The remedy against this bias is the introduction of "normative" calculation procedures. These procedures are derived such that the indicators are objective measures of a certain performance aspect. Although the resulting value may not always be interpretable as an absolute (in physical terms) measure for the observed performance, the approach is ideal for comparative studies. An even bigger advantage of a normatively declared PI is the fact that its value is directly related to the relevant set of building and client parameters.

Taking the reasoning from the previous section, a normative indicator is a measure that is defined through a simplified but indicative experiment. The simplification of the experiment allows the derivation of and aggregation over the output state of the experiment to be expressed as closed equations. These equations reside as spreadsheets or formulae in the toolkit, whereas the toolkit user is automatically prompted for all the needed building and client parameters.

In the case of thermal comfort performance, the choice of necessary PIs was driven by the following observations. Thermal discomfort occurs frequently for one of the following reasons: (1) the required average temperature is not maintained by the HVAC system; (2) the air supply system creates discomfort zones close to diffusers; (3) discomfort due to asymmetric exposure to colder façade elements; (4) cold drafts in the perimeter zone; (5) combining different dress codes and activities in one zone, creating too much comfort range diversity in the same zone; (6) inadequate zoning of the HVAC supply and control system; and (7) the inadequate capacity of the HVAC system under maximum load conditions. In actual situations the total discomfort will result from the superposition of all of these phenomena. Looking at it this way, it is not surprising that thermal discomfort is still complaint number one in office buildings! The novelty of the toolkit is its capacity to deal with all these phenomena separately, i.e. by introducing one or more PIs for each of the discomfort "agents," as indeed table 7.1 suggests. Each of the PIs is evaluated based on a simple experiment leading to a closed form calculation, some of which are based on existing normative calculation procedures. The Air Diffusion Performance Index (ADPI), for instance, was originally developed as a design guideline for selecting appropriate air supply devices. In our case it is used as PI1, a measure of temperature variation, air mixing and the presence of objectionable drafts.

To account for the asymmetric thermal radiation, which is one of most common reasons for discomfort in perimeter zones in offices, three PIs (PI2, PI3, PI4) are introduced in the toolkit. PI2, PI3 and PI4 are based on the PMV and the Predicted Percentage of Dissatisfied occupants (PPD). The value is normatively calculated from the glazing surface temperature and the corresponding Mean Radiant Temperature (MRT) of the façade. The calculation routine is based on the diurnal and seasonal hourly weather variations obtained from TMY2 data records (Typical Meteorological Years). The hourly glazing temperature is calculated by using a steady-state calculation method with information on the U-values of 48 different types of glazing systems provided by ASHRAE[19] and the corresponding mean radiant temperature is estimated in accordance with ASHRAE Standard 55.[20] The other PIs in the thermal comfort category are all developed along similar lines.

It should not come as a surprise to the reader that a normatively declared PI calculation has to be tested carefully in order to ascertain that it is indeed a good and monotonic indicator of the considered performance. Monotonic behavior is especially important as one needs to ascertain that an increase (in this case the optimal PI value is zero) in the PI value leads indeed to a monotonic increase in discomfort.

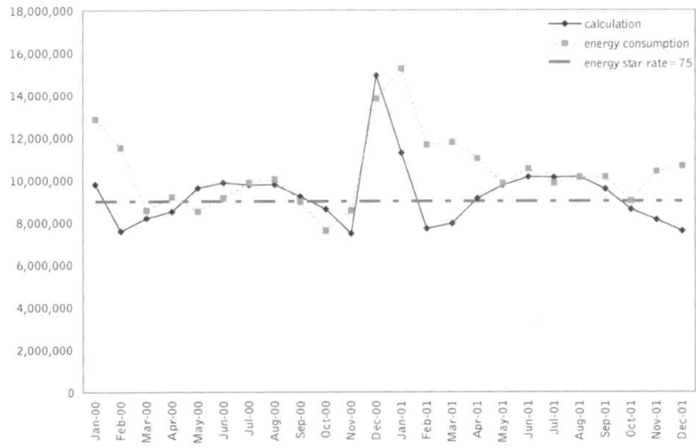

7.7
Comparison of the calculation and energy consumption (MJ)

7.6
Aerial view of the Sam Nunn Atlanta Federal Center (AFC), Atlanta, USA (1997–98), architect Kohn Pederson Fox Associates (the previously existing Rich Annex dates from 1924).

Of course one of the immediate benefits of a set of the introduced set of measures is its unequivocal use in statements of client expectations and its use in rapid design evaluation. Owing to the direct relationship to design and occupant parameters, one can directly inspect the influence of certain design choices.

The toolkit is currently benchmarked in ten office buildings in the south east of the United States. One case of the benchmarking in the area of energy performance is reported below.

As one of the targets of the energy PI benchmarking, the Sam Nunn Atlanta Federal Center (AFC) in Atlanta, Georgia, was chosen. The AFC (figure 7.6) is one of the largest federal office buildings on the east coast and provides office facilities for multiple federal agencies. The AFC consists of four connected buildings (a 24-story tower, a 6-story bridge, a 10-story mid-rise tower and a 6-story building). The total building is served by five chillers. The total floor area is about 148,000 m^2 and the total number of occupants is approximately 4,500.

For data gathering and survey, three visits were made to the building. First, information such as floor area, wall area, window area and U-values was calculated from available building records. Then, relevant system information, such as the HVAC system and its operation scheduling, HVAC fan powers (kW), etc., were obtained through the GSA facility manager.

After filling the toolkit data entry form with all relevant data items, the energy PIs show up as revealed in table 7.2.

Table 7.2
Breakdown of energy consumers in the Sam Nunn building.

Performance indicator	Energy consumer	Average in MJ (2000–01)	
PI1	Heating	8,553,494	7.6%
PI2	Fans for ventilation and air circulation	9,090,290	8.0%
PI3	Lighting	54,949,602	48.6%
PI4	Pumps	4,579,134	4.0%
PI5	Cooling	30,225,076	26.7%
PI6	Humidifying	4,006,742	3.5%
PI7	Domestic hot water	1,691,157	1.5%
	Total	**113,095,495**	**100.0%**

In order to make a comparison with real consumption data, the computations were carried out with the actual Atlanta climate for 2000 and 2001. Note that the normative calculations are based on the standard reference year for Atlanta, enabling a year to year comparison of the energy performance, not biased by the annual weather variations. For benchmarking purposes, we have to compare PI values with actual use. In the case of the energy PIs, this is warranted as the normative procedure gives reliable estimates of the expected real energy consumption.

A big advantage of the normative calculation is that it provides a breakdown of energy consumers, as shown in table 7.2. The outcome of the breakdown will help building owners and facility managers to have a sense of where the attention should be paid and where the budget should be allocated to improve energy efficiency.

Owing to the lack of energy sub-metering in the Sam Nunn building, the breakdown in table 7.1 could not be compared with metered data on-site. But it should be well understood that this (diagnostic use) is not the prime purpose of the PI. Rather, it defines the performance of the different energy consumers, enabling comparison across buildings of the energy consumed by each of the categories. It will enable the building owner to define the minimum performances for the lighting or cooling contingent on certain client and building types. This will lead to a much more refined "energy service management" based on the PI instruments than currently offered by consumption measurement (with or without sub-metering).

The differences between normative and actual total yearly energy consumption in the year of 2000 and 2001 are 2.2% and 15.4%. The deviations occur mostly during the winter season (figure 7.7). As explained before, the deviations between the normative expectation and the real energy consumption are attributable to occurrences that are not part of the "as designed" operation of the building.

Table 7.3
Obtainable annual energy savings calculation.

	From	To	Savings (MJ)	$/year
Orientation	South East–North West	South–North	1,373,109	16,877
Windows ratio	40%	30%	3,783,711	46,507
Lighting	Central on/off	Daylight switch	10,989,920	135,082
		Total	**16,020,313**	**196,913**

In this case the reasons for the deviations became obvious during an on-site inspection. The two main reasons were: (1) the full-time (including nighttime) operation of the HVAC system to prevent freezing (this was outside the use specification of the system and therefore not considered in the normative calculation); and (2) the use of personal heaters, mostly by workers in the perimeter zones of the office spaces (the thermostat settings of the "as designed" operation proved to be too low to provide sufficient levels of thermal comfort).

The AFC did not obtain Energy-Star labeling over the last two years. Such a label is given to buildings that are more energy efficient than 75% of their "peers," based on actual utility billing. The AFC currently ranks only in the 62 percentile on the Energy-Star office scale, and the energy bill in 2001 was about $1.6 million. In order for the building to obtain Energy-Star labeling, an 18% reduction of energy consumption would be required. By using the toolkit, a simple sensitivity study was easily made by changing several parameters, showing how the building's Energy-Star rating could be improved from 62 to 68, as shown in table 7.3. As exemplified here, the "normative" nature of the toolkit makes it readily usable for a sensitivity/feasibility study of: (1) new buildings in the design stage; or (2) existing buildings to be refurbished.

Note that even with a combination of design (orientation and windows ratio) and operation changes (lighting controls), the building would not obtain the expected Energy-Star rating of 75. In the current building only a drastic redesign of the lighting controls (including task lighting) could possibly accomplish this.

The ease with which this recommendation was generated, based on the use of the toolkit, is regarded as a strong endorsement for the development and expansion of the PI approach. The benchmarking of the other PIs shows similar encouraging results.

CONCLUSION

The introduction of PIs as unbiased measures of performance has great potential for enhancing the dialogue between stakeholders in the building delivery and management process. It has been shown how a system-theoretic approach may enable the systematic introduction of PIs. This way, a growing set of quantifiers of building performance are offered that could be infused in more rational decision-making between building parties, such as in the dialogues between client and architect, between architect and engineer, and between tenant and portfolio manager. The introduction of toolkits to "operationalize" the PIs is an important step in this infusion process.

A number of important research questions need to be addressed before large-scale toolkits will emerge:

1. Can a limited and distinct set of PIs be defined that cover a significant percentage of recurring dialogues?
2. Can the approach capture performance at increasing levels of granularity in accordance with design evolution or will the necessary number of PIs "explode"? The answer to this question will be determined largely by the establishment of a PI classification and the way it can be managed for different building systems on different levels of granularity.
3. Will rationalization lead to the capture of best practice in current building/engineering design and thus be able to act as a catalyst for re-engineering? The building performance analysis profession is gaining in maturity but lacks clear standards and accepted quality assurance methods. The diffusion of best practices could prove to be an important factor for this maturation and could foster the further development of performance measures.

NOTES

1 Lauri Koskela, *An exploration towards a production theory and its application to construction,* Dissertation, Espoo, Finland: VTT Building Technology, VTT Publications 408, 2000, p. 296.

2 G. C. Foliente, R. H. Leicester and L. Pham, "Development of the CIB Proactive Program on Performance-based Building Codes and Standards," *BCE Doc 98/232,* Australia: International Council for Research and Innovation in Building and Construction (CIB), 1998.

3 Jeffrey L. Beard, Edward C. Wundram and Michael C. Loulakis, *Design-Build: Planning Through Development,* New York: McGraw-Hill, 2001.

4 Godfried Augenbroe, "Performance aspects of building systems design," Graduate course syllabus, Georgia Tech, College of Architecture, 2004.

5 Greg Foliente, "Developments in Performance-Based Building Codes and Standards," in *Forest and Products Journal* 50: 7/8, 2000.

6 ASTM, "ASTM Standards on Whole Building Functionality and Serviceability," West Conshohocken, PA: ASTM, 2000.

7 Francoise Szigeti and Gerald Davis, "Matching People and Their Facilities: Using the ASTM/ANSI Standards on Whole Building Functionality and Serviceability," in *CIB World Building Congress,* April 2001, Wellington, New Zealand, 2001.

8 Ibid.

9 W. Gielingh, "General Reference Model for AEC Product Data," *ISO TC184/SC4 Document Number 3.2.2.3.* Delft, Netherlands: ISO, 1987.

10 H. A. J. Ridder, *Organization of Complex Problem Solving Processes,* PhD Thesis, TU Delft, Delft, 1996.

11 J. G. Gross, "Developments in the Application of the Performance Concept in Building" in Rachel Becker and Monica Paciuk (eds), *3rd International Symposium: Applications of the Performance Concept in Building* 1, Tel Aviv, Israel: The National Building Research Institute, 1996, pp. I-1 to I-11.

12 Foliente *et al.* (1998), op. cit.

13 D. J. Vanier, M. A. Lacasse and A. Parsons, "Modeling of User Requirements Using Product Modeling" in Rachel Becker and Monica Paciuk (eds), *3rd International Symposium: Applications of the Performance Concept in Building* 2, Tel Aviv, Israel: The National Building Research Institute, 1996, pp. 6-73 to 6-82.

14 Ibid.

15 P. O. Fanger, *Thermal Comfort,* Copenhagen, Denmark: Danish Technical Press, 1970.

16 ST&M, *Serviceability Tools and Methods, ASTM standard* in Gerald Davis and Francoise Szigetti, ASTM, E. 7, 2001.

17 US Green Building Council (USGBC), *LEED Rating System,* version 2.0, 2001.

18 Godfried Augenbroe, Ali Malkawi and Pieter de Wilde, "A Performance-based Language for Design Analysis Dialogues" in *Journal of Architectural and Planning Research,* 2004 (in press).

19 ASHRAE, *ASHRAE Fundamentals,* Atlanta, GA: ASHRAE, 2001.

20 ASHRAE, "Thermal environmental conditions for human occupancy," *ANSI/ASHRAE Standard 55-1992,* Atlanta, GA: ASHRAE, 1992.

8

ENGINEERING COMPLEXITY: PERFORMANCE-BASED DESIGN IN USE

CRAIG SCHWITTER

8

ENGINEERING COMPLEXITY: PERFORMANCE-BASED DESIGN IN USE

CRAIG SCHWITTER

Pressure for speed, economics and reliability in commercial building and infrastructure design has traditionally led engineers to rely on standard, code-based prescriptive approaches to solutions for structural and environmental systems. But as architectural solutions become increasingly complex, in part due to the power of expanding arsenals of architectural software and the intense spread of architectural information and construction technologies on an international basis, engineers must continue to offer solutions that enable the boundaries of building design to expand and flourish.

New developments in performance-based design offer a rapid and technical approach to unique systems analysis — methods that in the hands of design engineers, applied at the right time in the design process, can develop quality building design solutions well beyond the typical and prevalent prescriptive practices in use today. Building and infrastructure engineers are seeing rapid changes in the tools we use — changes that can lead to enhanced design capabilities and more creative thinking under the right circumstances. The architectural community is keenly aware of these developments and is beginning to capitalize on the emergence of new computational tools in building design. Structural engineering, HVAC engineering, fire and life safety design, and construction engineering are all poised to capitalize on the emergence of new computational tools in the future — a future for architecture that will have different market demands, design codes and technical challenges than we can imagine today.

CHANGE AND INNOVATION

Engineers may often resist it, but design is really all about change. While the evolution of building practice relies on the predictability and constant nature of science and materials, it also requires us to constantly develop new design approaches for an ever evolving architectural and infrastructure landscape. But how do we as engineers change and innovate? With a culture that is littered with litigation, and one that is slow to adopt change, the construction industry is hardly the place to look for allies in this effort.

However, change is inevitable — a part of fundamental human development. It is a simple fact that the building engineering solutions of today will become outdated over time. As engineers, we can contribute significantly to the effort for change but often we do not. Engineers are more typically reliant on "rule of thumb" design or processes that are familiar and proven — an understandable pattern, especially as an engineer's view must be conservative at times. But what about intellectual curiosity and the creativity of engineering design? What about using "what if" more often? A stubborn attitude to change is the engineering profession's heaviest albatross, and yet, at its most fundamental practice, engineering design, or the application of science, has potentially the most to offer in the evolution of the built form.

Even as we strive for change, we must look back as much as we look forward. Prescriptive design practices, established over many years and with a sound technological basis, need not be thrown out to make way for new ways of thinking. Instead, we must look to build on prescriptive methods and, with the use of rapidly evolving computational tools, begin to explore more fully the world of performance-based design in engineering. Examples abound in performance-based engineering design, many making use of the most sophisticated tools available today, such as computational fluid dynamics (CFD) modeling to computer-aided manufacturing processes. These are discussed in more detail later, but first we need to establish more directly the distinction between performance and prescriptive approaches to engineering design.

PRESCRIPTIVE DESIGN AND OPTIMIZATION VERSUS PERFORMANCE-BASED APPROACH

A prescriptive approach is the most commonly adopted approach in engineering design. Prescription implies that there is a set of rules that need to be followed, rules normally outlined by a code or a design guide that is based on previously developed empirical and scientific knowledge. A simple example is the design of a beam from reinforced concrete. The design of reinforced concrete structures, prior to the early 1900s, required a leap of faith or a belief in the many opinions and different calculation methods available at the time. Today, the design of a concrete beam follows standard rules of practice, so much so that we can look to tables for most element sizing. This is not to be seen as detrimental as it allows us, as building engineers, to process certain aspects of design quickly. We are standing on the shoulders of those who have come before us, and using this knowledge is important to speeding up the process of design.

But prescriptive design takes on larger importance when one views it with respect to the larger scale design of systems, such as structural or mechanical systems for buildings. Often due to economic and time pressures, building form is limited in complexity in order to simplify large-scale inputs for engineers — simplifications that manifest themselves in ensuring that the design and the spatial configuration have been seen before and that the design process is predictable. Lessons learned or technologies developed from the previous design of a similar building type are utilized to produce quick and normally economic designs. The design process can thus be a linear one with the rest of the design and construction team, since the design outputs are already well understood at the outset and risk is minimized. In commercial practice, this approach is dominant as it is more time efficient and less encumbered by the unknown. For much of the building practice, the predictability of known inputs and outputs is welcome.

It is important to recognize that the engineering efforts in this type of prescriptive design work are concentrated mainly in the optimization of the system. Design evolution can often be categorized in quantities, such as pounds of steel or tons of cooling — traditional markers of "efficiency." Commercially available engineering design tools today are predominantly aimed at this optimization process.

Engineering design is often hijacked by the optimization process, and while this can be an important part of engineering design, it should not be misconstrued as more significant advances or quantum creative leaps in engineering design — advances that are in large part due to thinking outside of the established solution and its associated parameters. Also, the quantities being optimized, such as pounds of steel for structural systems or cubic feet per minute (CFM) of air supply for HVAC systems, can often start to change as market conditions evolve, leaving established and previously optimized solutions behind the curve.

It is inevitable that business and the human conditions combine at certain points to challenge past performance and to question previously held convictions of how to approach a design. A prescriptive approach may provide reduced risk, but it can also lead to reduced gains. A performance-based approach may be more tailored to use in a particular project — one where the design problem cannot be simply categorized, or a solution from the past be readily adopted. Performance-based design offers a process that relies more fully on an engineer's training in creative problem-solving and applying first principles for design. It does not preclude the use of prescriptive areas of design where they may be applicable, either at a component or systems level. But in a performance-based design process, the design inputs have to be carefully developed and meticulously understood. Their effect on the design is critical to the development of an innovative product. The process of design feedback is also critical to the success in performance design. Finally, a design output cannot be prejudged or biased. The process must be relied upon to properly assess inputs to develop and shape an unknown outcome. In this way, emerging computational tools are critical to the process, not as tools for optimization but as technological guideposts for solving complex problems.

It is a fact that prescriptive processes are utilized constantly in building engineering, as our processes for performance-based design are in their infancy. This is for good reason, as performance-based design eliminates several hurdles, the greatest of which is the requirement for integrated design thinking for all members of a design team. For performance-based design to work effectively, it relies on well-communicated feedback loops between different members of a design team. This, in turn, creates a process that is more non-linear than a standard prescriptive approach — a design hysteresis loop than can often be frustrating and time consuming. Again, the use of computational tools can assist and improve this process, but

8.2
Darien Lake Pavilion: tensile fabric structure roof, internal view. Structures are created from patterned fabric panels into large three-dimensional surfaces.

8.1
Tensile fabric structure roof, exterior view. Non-linear finite element programs allow form development and the analysis of non-geometrical shapes. Darien Lake Pavilion, Buffalo, New York (1996), architect FTL Design Engineering Studio.

the real challenge is in the management of the design process and in ensuring that design iterations conclude rapidly. An integrated approach to engineering design — one that adapts quickly to other disciplines and is rooted in a fundamental understanding of a variety of building systems — is essential to the success of a performance-based approach.

THE QUANTITATIVE BIAS TO DESIGN: EVOLVING TOOLS FOR PERFORMANCE-BASED DESIGN

If engineering design processes are often behind the technological curve, today's engineering design tools, on the contrary, are increasingly sophisticated and advanced, opening up many opportunities well beyond established prescriptive design methods. There are several obvious reasons for this current development in the engineering profession. Certainly one is that building and infrastructure engineering are benefiting from crossover technology from other disciplines, such as aeronautical engineering, where computational modeling is significantly more advanced. While computer-aided design (CAD) use is widespread, the use of higher end computational tools is only now becoming more typical in the offices of building engineers. Second, the rapid increase in computing power, coupled with the strong and ever-developing software industry, has fueled the emergence of sophisticated computational tools. The evolution of computing power will march on and we should be aware of what changes this will bring about in our ability to accurately model and simulate for building engineering.

Finally, and most importantly, is an overall quantitative bias to engineering design, which is emphasized both in academic institutions and by society as a whole. The bias is felt often in areas of "softer" engineering, or where design is emphasized over scientific development. The drive for building engineers to become scientists is real, and the reliance on sophisticated computational tools is a byproduct of this hidden sociological emphasis. This bias has a positive side for the design engineer, of course, and computational tools are used primarily today to provide "proof" of previously recommended designs. However, computational power and crude design processes have in the past hampered efforts to use computational tools as real design drivers. The challenge to today's engineers involves seeking real ways of moving computational tools from simply being a means of proving design ideas to being integral parts of the design process, parts that can provide design input quickly and iteratively. This is essential to bringing performance-based design solutions to bear on problems. Areas of current interest to Buro Happold in performance-based design include structural engineering, advanced HVAC design, CAD/CAM (computer-aided manufacturing) processes and building simulation.

PERFORMANCE-BASED DESIGN EXAMPLES

In the structural engineering field, finite element programs have been established as the primary computational tool for the past thirty years. Recent advancements in graphic and CAD modeling software have aided the development of more aggressive structural forms, and fortunately commercially available finite element programs have kept up the pace. For more specialized areas of structural engineering, many firms have developed performance-based software suited for specialized uses. Buro Happold developed a non-linear finite element program, called TENSYL, for the original purpose of analyzing cable nets and tensile structures. Pure tensile surface structures are a good example of performance-based structural design (figures 8.1 and 8.2). The forms for these structures are not designed, in as much as they are form found from boundary conditions with surfaces calculated between these boundary inputs. The surfaces themselves are non-geometrical, as they do not follow any mathematical

8.3
Darien Lake Pavilion: computer-aided fabrication of patterned fabric roof panels.

8.4
Stuttgart 21, Stuttgart, Germany (2001–), architect Christoph Ingenhoven Overdiek Architekten GmbH und Co. New railway station roof made of concrete shells formed with tensile membrane models.

8.5
Stuttgart 21: computer rendering showing staggered individual shell units.

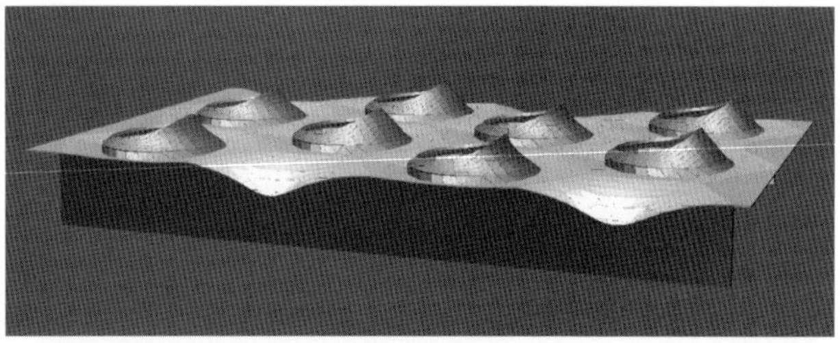

8.6
Stuttgart 21: physical models for the tensile forms were recreated computationally prior to "hardening" into stiff shell elements for structural analysis.

formulae. Structures such as this have only been economically achievable with the assistance of computational models for load analysis and surface patterning for fabrication (figure 8.3).

More recently, TENSYL has evolved into a powerful tool for amorphous structural and geometrical analysis, as shown in its use on the Stuttgart 21 project (2001–) where intricate physical models constructed by the architectural team were reconstructed computationally for the first stage of analysis for the project (figures 8.4–8.6). In this process, tensile mesh models were form-found to develop compressive shell surfaces that make up the roof of a new major train station in Stuttgart, Germany. The shell surfaces were developed with openings to allow light penetration from the upper level park down into the new track levels.

Other recent efforts to investigate economical construction of amorphous forms have used performance-based design. The use of principal curvatures is a method to economically "map" a curving surface into flat planes, avoiding the use of costly curved glass or other component-cladding materials. Buro Happold again developed a specific software routine to use this method effectively on a range of surface structure projects. One example is the design of a recent 20,000 square foot glass canopy project (Roppongi Canopy, Tokyo, Japan, 2004–; figures 8.7 and 8.8). In this example, the structural system was designed to facilitate the principal curvature lines for achieving flat panels over the curving structure surface.

8.7 (left and right)
Roppongi Canopy, Tokyo, Japan (2004–), architect Skidmore Owings and Merrill (SOM). The initial model where straight lines are laid on a curving surface resulting in curved glass cladding panels throughout.

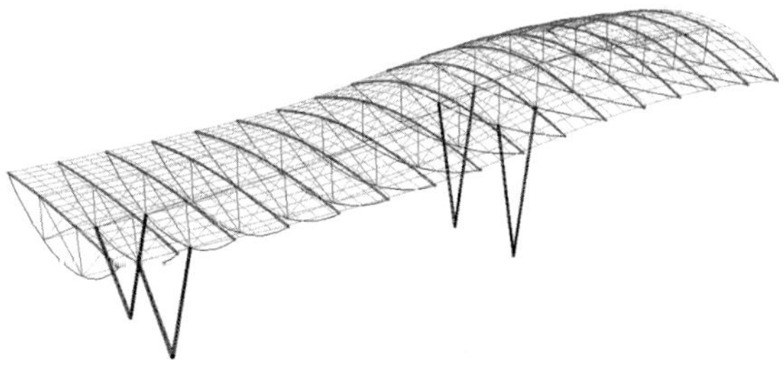

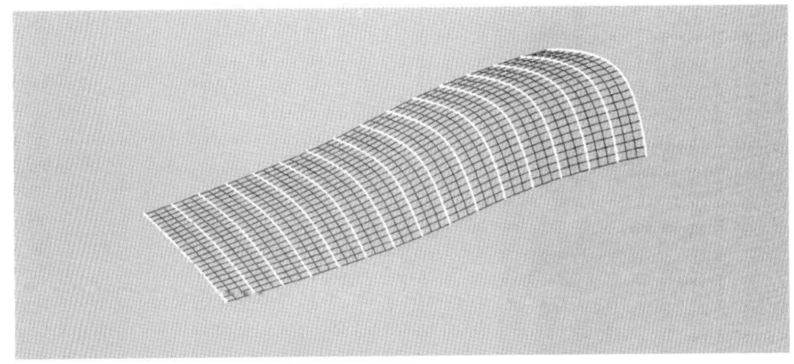

8.8 (left and right)
Roppongi Canopy: the developed surface model with principal curvatures providing the continuous curving surface with flat trapezoidal glass panels.

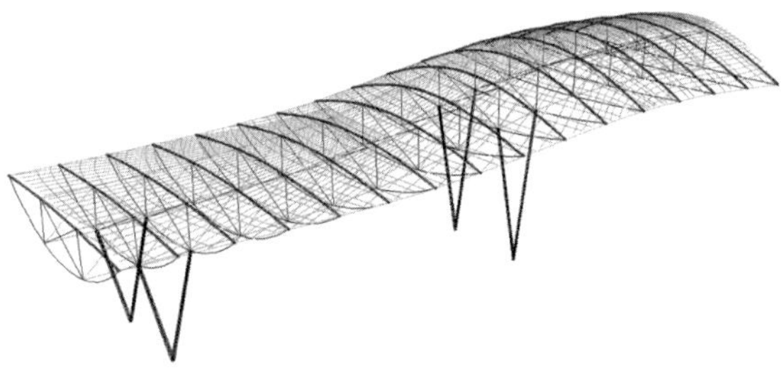

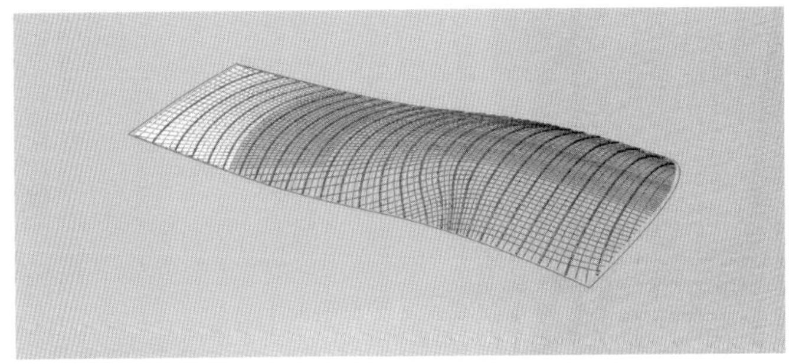

8.9
Japanese Pavilion, Hanover, Germany (2000), architect Shigeru Ban Architects. Exterior view showing cardboard gridshell under construction.

Materials offer some of the most exciting advances in performance-based design, as design processes and projects can be developed specifically to suit the material's characteristics. The use of standard paper or cardboard in construction has been used by Buro Happold on several projects, including the Japanese Pavilion in Hanover, Germany (2000). While it has compressive strength along the lines of wood, it is limited in application by its overall stiffness and a propensity to creep with time. However, as a completely recyclable material and one available inexpensively throughout the world as spiral wound tubes, it is an attractive material for temporary structures. A performance-based approach was necessary as the cardboard material has no established prescriptive design guidelines. Through extensive testing of the strength and stiffness limits of the material (figures 8.9–8.11), Buro Happold was able to develop suitable analysis and design procedures for the material and have the confidence to build with it on a large scale.

8.10 (right)
Japanese Pavilion: interior finished space with paper cladding and cardboard tube structural system.

8.11 (far right)
Japanese Pavilion: the extensive testing of cardboard tubes was required in order to establish design guidelines for the use of the material in construction.

8.12

The Experimental Media and Performing Arts Center, Rensselaer Polytechnic Institute, Troy, USA (2003–), architect Nicholas Grimshaw and Partners. The CFD model of the inside of the concert hall, showing relative air velocities in the displacement ventilation system.

8.13

The University of Michigan Biomedical Research Building, Ann Arbor, USA (2001–), architect Polshek Partnership Architects. The view of the CFD model for the double façade cladding wall.

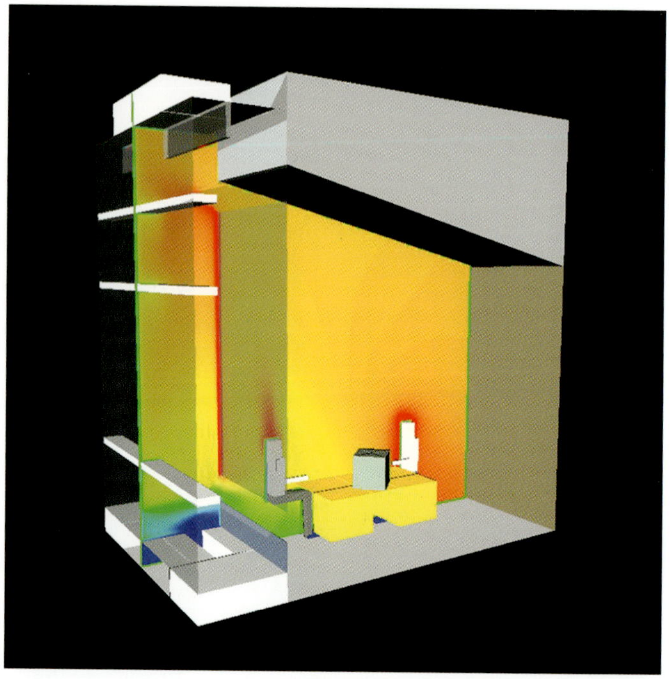

8.14

The University of Michigan Biomedical Research Building: various CFD models for lobby winter and summer environmental modeling.

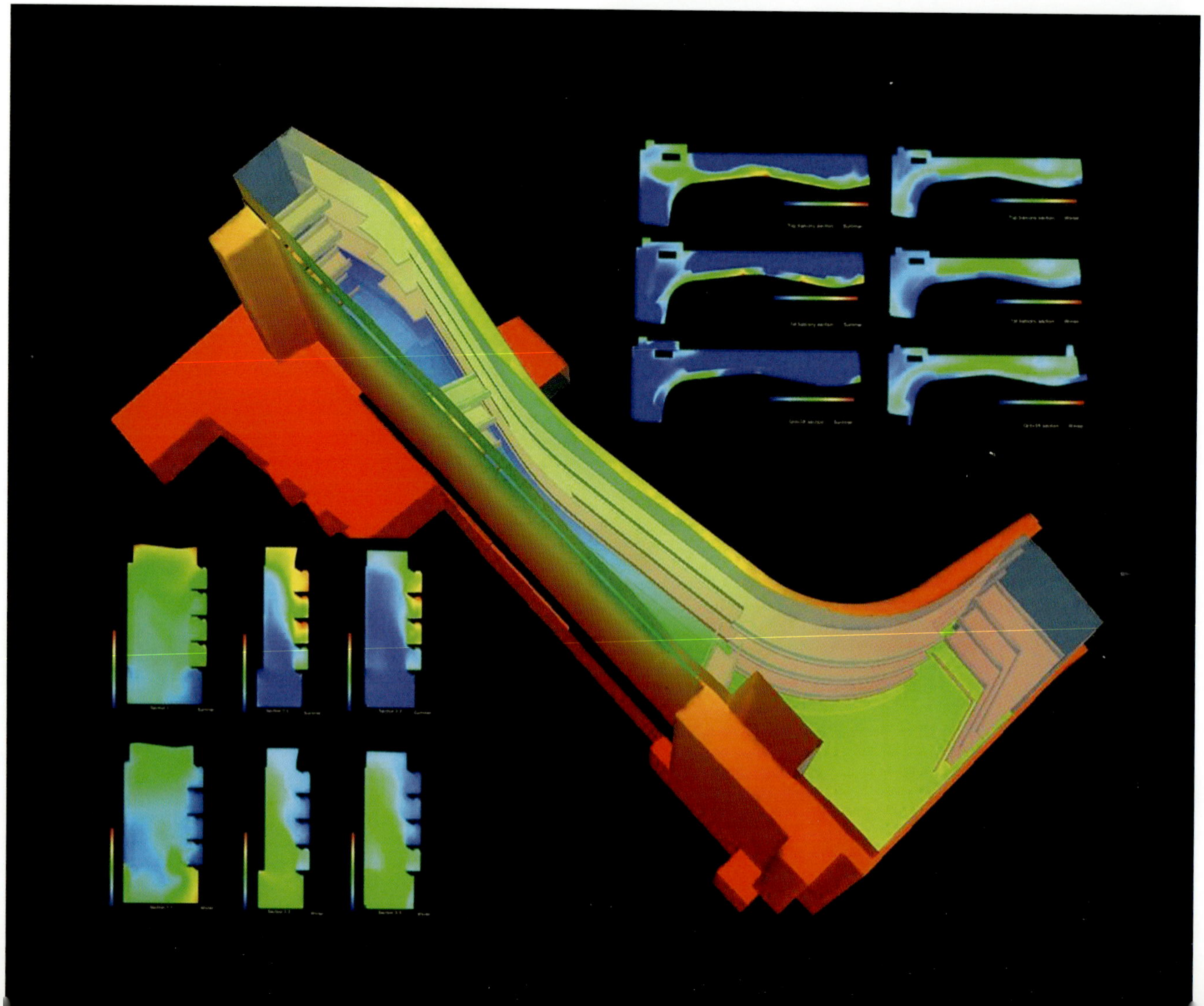

8.15
BBC White City, London (2002–), architect Allies and Morrison. The CFD model of external air pollution on site.

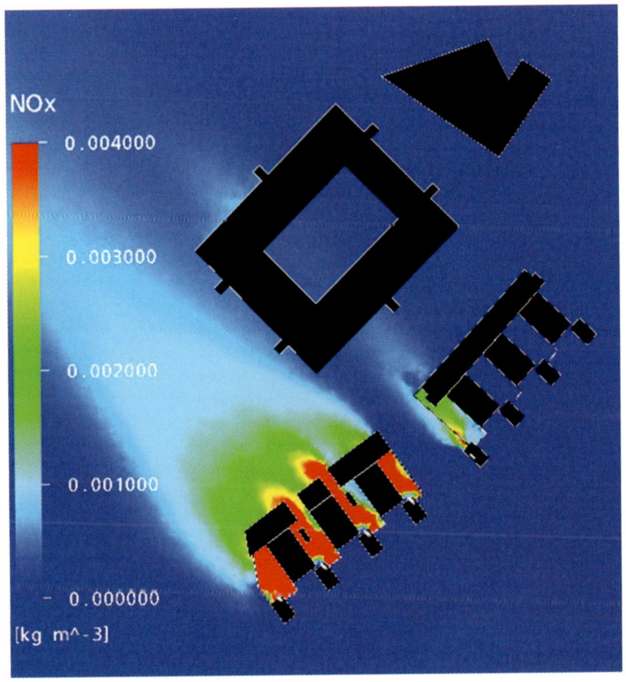

8.16
BBC White City: the computational model indicating levels of light pollution.

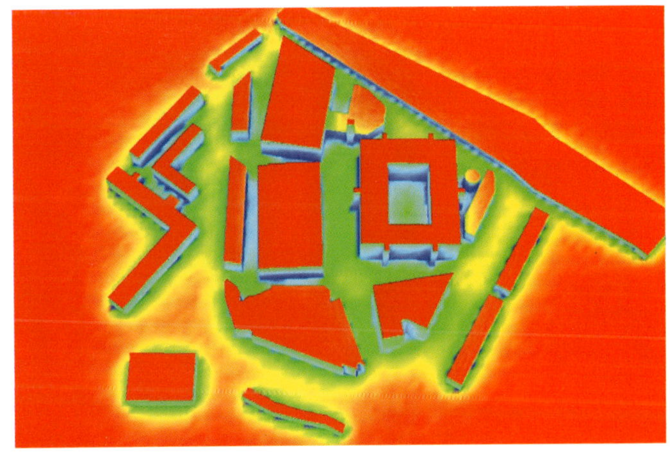

Mechanical engineering and, in particular, HVAC design have long been resistant to change and have been slow to adopt computational tools. Significant advances in recent years have made CFD modeling an important part of building engineering, in particular the development of HVAC systems for spaces where conventional, prescriptive approaches are not appropriate. Figure 8.12 shows a CFD model of the inside of a concert hall for the new Rensselaer Polytechnic Institute's Experimental Media and Performing Arts Center (Troy, USA, 2003–). The model was required due to the proposed displacement ventilation solution, where low velocity air is introduced through plenums below the seating areas. The CFD models were used to refine the HVAC strategy and to ensure that the internal comfort was maximized for the system. Other examples include a large lobby space and double façade cavity wall system for a biomedical research facility located at the University of Michigan (Ann Arbor, USA, 2001–). CFD modeling was used extensively to underpin the environmental strategies put forward by the architectural design team for the project (figures 8.13 and 8.14).

In addition to CFD, energy modeling and thermal and lighting simulation are quickly becoming essential for low energy and high-performance building system design. Environmental effects, at both a micro and macro level, can now be tested and are able to shape and inform critical design decisions. Figure 8.15 shows pollution dispersion modeling for the BBC White City project, London (2002–). The CFD simulation critically reviewed the pollution generated by the campus power generation equipment. Additional simulation modeling (figure 8.16) shows the effect of light pollution on the site.

CAM tools have also been long associated with advanced fabrication solutions to building components. Again, recent advances, primarily in architectural modeling tools and the proposals of aggressive and non-geometrical forms, have quickly been pushing CAD/CAM technologies forward. A new generation of architects is now becoming familiar with rapid prototyping, 3D printing and a whole host of new CAD/CAM techniques that were traditionally developed for industrial design applications, primarily in the auto industry. These tools affect not only the product but the mindset of the new designers, and it is clear that CAD/CAM processes for large-scale buildings are gaining significant momentum. Engineering designers who are able to not only use these new software programs but, more importantly, use these

8.17
Gateshead Music Centre, Gateshead, UK (2003), architect Foster and Partners. The roof structure provides clear span cover for the music center. Gateshead Music Canopy, Gateshead, UK (2003), architect Foster and Partners.

8.18
Gateshead Music Centre's roof: steel rib structure under construction.

8.19
Gateshead Music Centre: parametric modeling for roof structure to provide economy in constant circular radius primary and secondary members.

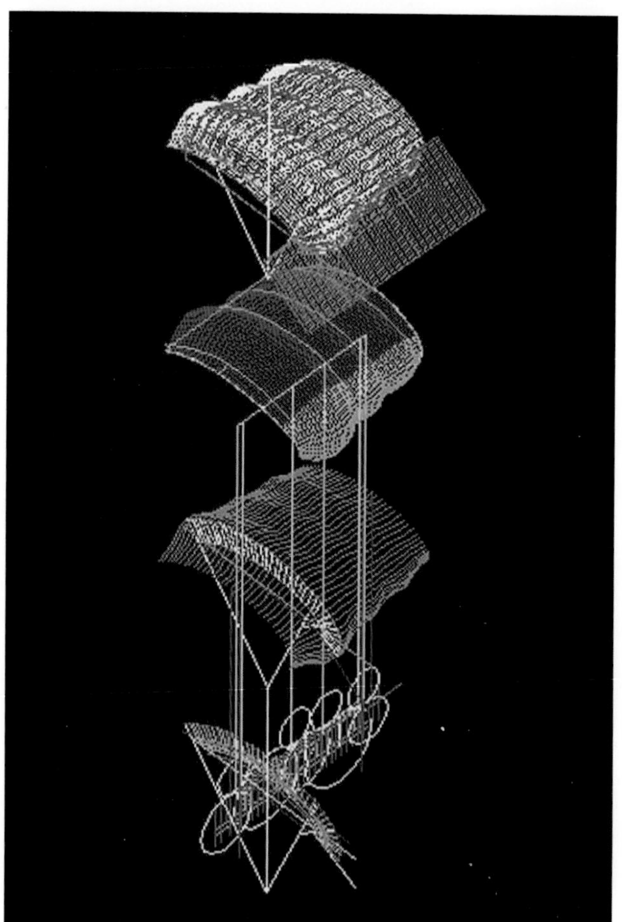

fabrication processes to inform their designs, will push the boundaries of performance-based building design in the future. For example, figures 8.17–8.19 show the Gateshead Music Centre's roof (2003), a structure developed parametrically to maximize the use of consistently radial primary and secondary steel framing. The British Museum's roof (2001) also illustrates an example where CAM made the project economical and achievable (figures 8.20–8.23).

Finally, simulation tools involving human behavior are now becoming useful in developing and testing new design strategies for public spaces. People movement software simulates individuals and response times to either life safety events or simply the morning rush hour: these are then calculated with respect to the building design and layout. Figures 8.24 and 8.25 show the use of people flow modeling for crowd control around the new Arsenal Stadium project (2001–). These simulation tools provide a great leap forward into the next generation of performance-based design.

What may be most encouraging about today's engineering design tools is that we are only at version 1.0. Computational power, in accordance with Moore's law, will continue to drive what engineers can do and in what time frame. Shorter periods for computational modeling will continue to allow more design feedback loops and will hopefully lead to more integrated design for building and infrastructural teams.

8.20
Great Court, British Museum, London (2001), architect Foster and Partners. Interior view showing the curving glass roof over the central courtyard.

8.21
The British Museum: overhead view.

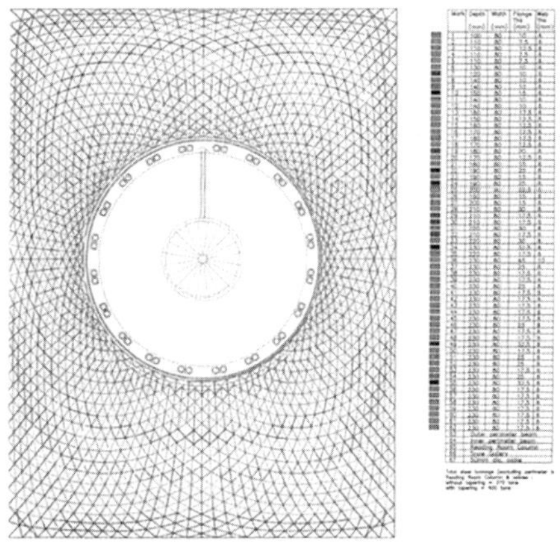

8.22
British Museum: drawing of differing element sizes for roof structure.

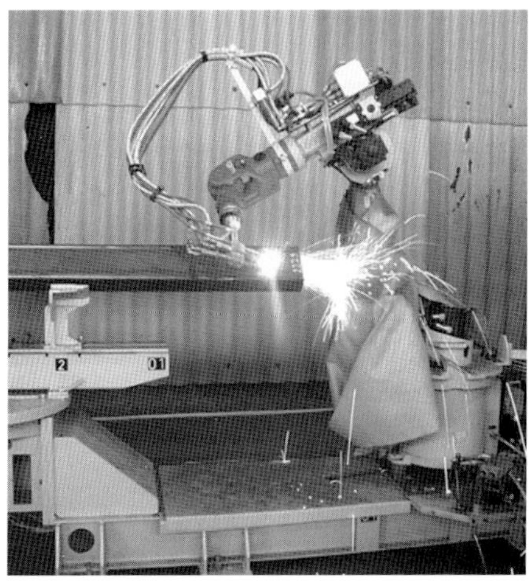

8.23
The British Museum: the CAM of all nodes allowed a large number of precise connections to be fabricated economically.

8.24
Arsenal Stadium, London (2001–), architect HOK Sport. People flow modeling is increasingly being used for assessing safety and comfort concerns at public venues.

8.25
Arsenal Stadium: crowd flow modeling indicating congestion points created by the large exodus of pedestrians after a game.

SUMMARY

Computational tools are clearly gaining ground in a variety of uses and their practical applications in many areas of building engineering design are helping to push the boundaries of quality and complexity in architecture and infrastructure. The challenge in the effective use of these tools will be how well they are adapted to specific problems, and how well they are integrated into the design process, a process that today can be free to be more performance-oriented than a more traditional prescriptive approach.

But let us not forget that the best computational tools do not offer solutions for design. They merely hint at this, and it is the responsibility of the designer to ensure quality and coherence in the design process.

Finally, it should be evident that today's performance-based approach may well be tomorrow's prescriptive recipe for a solution. The only constant in our world is change. And as engineers we need not resist this critical process of improvement of the built environment, but we must seek to think outside of the box for solutions for tomorrow's problems — solutions that ultimately ensure the highest design quality for our clients.

9

ENGINEERING IN A PERFORMATIVE ENVIRONMENT

JEAN-FRANCOIS BLASSEL

9

ENGINEERING IN A PERFORMATIVE ENVIRONMENT

JEAN-FRANCOIS BLASSEL

RFR's work on architectural structures and building envelopes has positioned the firm at a threshold that has long separated architecture and technology. That position has fueled the firm's interest in detecting a potential convergence between the two disciplines. Could it come now from the emergence of new modes of practice in both fields?

The practice of engineering has been moving from concepts and ideas towards concrete objects. Today, however, the analytical process tends to be displaced by modeling. Complex simulations of complete construction are increasingly being used to replace the rational process of unfolding a situation or a problem into a collection of individual, simple parts for eventual recomposition.

The mastery of the analysis is being displaced by the definition of the criteria — architectural as well as technical — that will segregate a successful experiment from an unsuccessful one. The criteria are not always quantifiable but all can be considered as part of an economy — monetary (always a powerful determinant in engineering) but also an economy of materials, of means, and even an economy of time.

CHANGES AND CONVERGENCES

The work of RFR revolves around architectural structures and building envelopes. We often find ourselves operating in a fuzzy zone where architecture and engineering overlap. This overlap explores a potential convergence between the two disciplines, and examines the ways in which the work of architects and engineers is being transformed by this proximity.

Architects who understand and accept the formidable power and responsibility that comes with modern technology have a great advantage in establishing their own methods for designing buildings for today — buildings that respond ethically to the current situation — sustainable, graceful, solid and economical. Very few teams have, so far, managed to transcribe this understanding into their practice. It is not only the architect, however, whose role is being transformed. Engineers, the traditional bearers of technology to the design process, also see their role profoundly transformed. Individual engineering specialties cannot work neatly in isolation anymore, ignoring each other. Engineers, as much designers as the architects, develop a capacity to work collectively and to take part in a project laterally. They must be not only highly competent in their often narrow field but also be able to replace their problems in a more global vision of the project or of the environment. Thinking about a project as a continuous transformation of the environment requires specialization and openness, invention and integration, and no blind spots. The emergence of a type of practice based on integration will hopefully liquidate the measure of technological performance by isolated quantifiable criteria. The criteria that spring from this new situation are much less definable and the engineering more complex. Borders are blurred: engineers tackle questions of architecture and offer solutions to make architecture.

Engineering has been described as an analytical process where one moves from concepts to a technical solution. This abstract process of analysis is being rapidly displaced by simulation, profoundly changing the way we work. Problem solving by analytical processes is being replaced by more complex simulations of complete built environments, from structure to cities. Of course, no tool is capable of an exhaustive simulation — but the mastery of the analysis is being displaced by the definition of the criteria. Choosing the right set of criteria by which to appraise the simulated projects becomes the core of the design: you design but at the same time you try to understand that which you are designing. These criteria, architectural as well as technical, create a dichotomy: they separate the successful experiments or models from the unsuccessful ones.

Changes can also be felt in the influence of economy on buildings. Monetary economy has always been a powerful determinant in engineering but there have always been other forms of economy, measured in different terms: there is an economy of materials, an economy of means, and an economy of time, for instance. If the nineteenth century was concerned with the economy of materials and matter, and the twentieth century could be best described by the search for an economy of labor and ultimately an economy of time, then the economy of the building industry in the twenty-first century will be an economy of *energy,* in all its forms — not only the drain on resources induced by the design during its useful life but also what it embodied at construction and, of course, what happens after the design is no longer useful. These two aspects of the changing conditions in which we design seem to influence the way in which buildings are made on at least two levels — where technology and architecture converge.

Technology influences, more directly than ever, the external morphology of buildings: directly, since the building's skin is the locus of exchanges of matter and energy between the building and its environment, and indirectly, since modeling tools allow almost surreal explorations of form. More fundamentally, buildings are technical objects — extensions of our bodies that allow us to do a number of things that we cannot do naturally.

Buildings are evolving in the manner of other classes of technical objects. The research of the evolution of the modern technical object generally reveals, at its genesis, a mode of composition based on juxtaposition and addition — the joining of autonomous parts; then, as technical objects become more sophisticated, their components tend to condense. In the case of the automobile, for example, the engine that was once sitting passively in the chassis plays a role in the rigidity of the frame of today's car. It is no longer independent from the body or the windshield. When applied to buildings, this design philosophy is manifested by the activation of building components. Elements or components of a building play multiple and complementary roles in order to increase the building component's physical usefulness beyond its initial destination.

The four projects that follow, one completed and the other three work in progress, illustrate these introductory remarks.

AVIGNON TGV TRAIN STATION

The TGV[1] train station in Avignon, France, opened in 2001. A multiplicity of criteria had to be considered and arbitrated during its design with the in-house architectural department of the SNCF, the French railway authority.

The station sits at the confluence between the Rhone and Durance rivers. It is a flat, hot, and windy site in the south of France. A very strong wind blows almost one day out of four. High rows of cypress and poplar trees protect the houses and fields in the surrounding area.

The elevated track and platforms increase the exposure of waiting passengers to the harsh natural elements, and the station is designed first as a shelter from the sun and wind. It works like an airport boarding lounge where one waits for the approaching train behind glass. The trackside of the building faces roughly north and the entrance side faces south. Despite its symmetry, the building section reflects the differences, of climate and use, between the two sides. Less space is needed at the end than at the center of the building, so the building tapers towards the end.

A large number of parametric studies, looking for a balance between repetition and tapering form, led us to choose a rotationally repetitive geometry. The building space is bounded by a small volume resulting from the intersection of two tori (figure 9.1), surfaces more easily described than the curves they generate where they meet. The two tori, each with its set of structural axes, generate their own structure (figure 9.2). The different qualities of the two opposite skins inform not only the surface but also the depth of the structure. Although it is shaped like a gothic arch (figure 9.3), the climatic asymmetry of the skin tells the structure what it wants to be. On the south side, the opaque skin shields the building against the sun and is very thick and stiff, although relatively light because of the high risk of earthquakes. Its construction relies on an unconventional use of conventional techniques: open section steel work, steel decking, insulation, waterproofing and cladding with a glass cement composite panel. The other side is totally different. It is as much driven by transparency as the opposite side is driven by opacity. Light and views are key.

9.1
The toroidal geometry of the TGV train station in Avignon, France (1998–2001), architect J. M. Duthilleul, engineers RFR Consulting Engineers, Paris.

The architecture of the insulating glass panels, deduced from the ridgeline for optical reasons, loses two key benefits offered by the underlying toroidal geometry: flatness and repetition. Instead, the panels are warped and unique. Their supporting structure seeks its slenderness — and transparency — in the adjacent building elements. It borrows its rigidity from the structure of the south side and from the concrete sub-level. The glass composition combines metallic coatings and ceramic frit of varying density to achieve optimal thermal performance as well as maximum transparency (figure 9.4).

The search for transparency led to a reconsideration of thermal standards. Contemporary energy codes severely restrict the amount of glass in a building envelope. Their strict application would have considerably hindered the design. However, in a train station, where people wear coats in winter, the interior temperature can drop to 14°C without creating discomfort. The building complies with energy codes directly and not by the installation of a standardized predetermined amount of insulation.

Optimal building performance comes from looking at things in a certain way. One gleans most of the necessary information from the singularities of the site and the future use, and not from some abstract and generic way of thinking. It is very difficult to attach a single criterion to a particular piece or complex of pieces, be it structure, skin or any of the building elements. They all interact with each other and one cannot select and decide to focus only on one aspect of performance to determine what the design of the building should be.

9.2
TGV station: structure under construction.

9.3
TGV station: typical section.

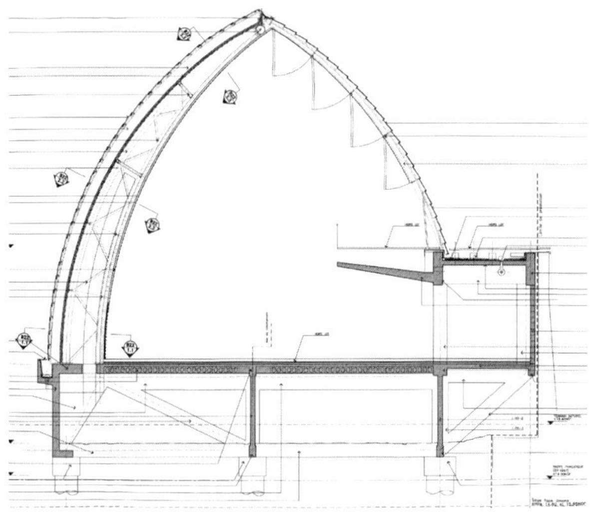

9.4
Interior view
of the train
station in
Avignon.

9.5
Bercy footbridge spanning between the Park of Bercy and the French National Library in Paris (1998–2005), architect Feichtinger Architects, engineer RFR Consulting Engineers.

9.6
Bercy footbridge: a composite computer rendering showing the bridge's multiple pathways.

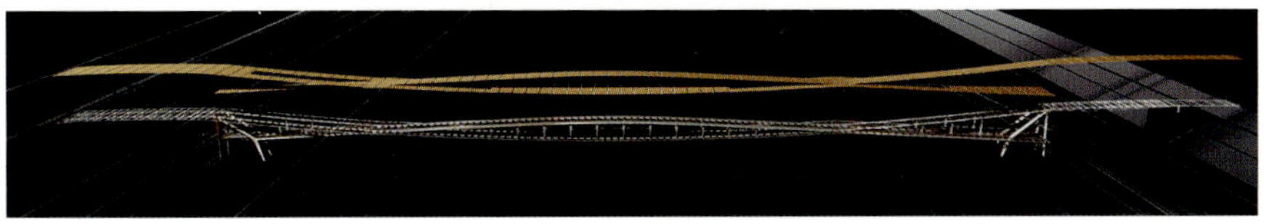

9.7
Bercy footbridge: the opposite curvatures of the bridge's pathways.

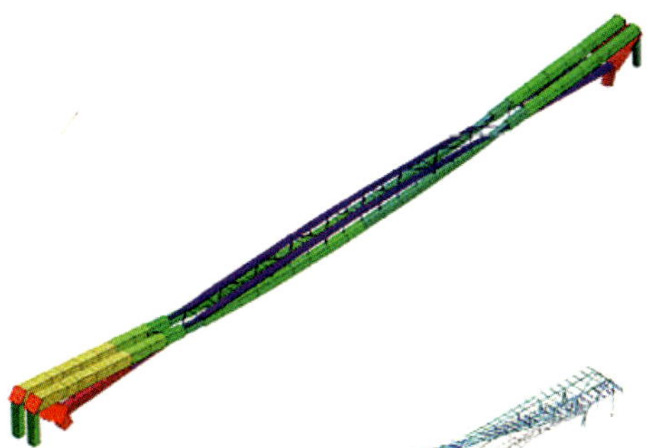

9.8
Bercy footbridge: finite elements analysis of structural stresses.

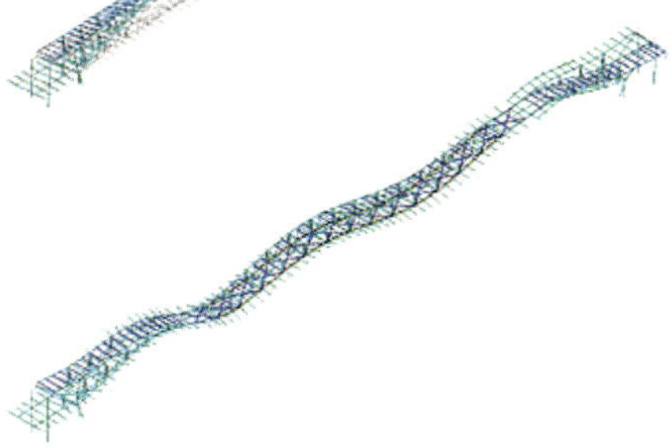

9.9a–b
Bercy footbridge: simulations of structural behavior.

BERCY FOOTBRIDGE

The design of a bridge is very heavily influenced by questions of spanning, but even with very long spans, other things inflect the design. To say that a structure is exclusively dictated by structural rationality is misleading.

The Bercy footbridge[2] spans the River Seine between the Park of Bercy and the French National Library in Paris. The structure is made to cross the river, but it is also made to link places in the city so that the paths in the city inform the way the structure is built (figure 9.5). Rather than building a span and then finding routes at the end, structure and path merge. Here, the bridge itself provides paths, not only across the water but also between the low banks of the river and the high level of the esplanade of the bibliothèque on the Left Bank and the park on the Right Bank (figure 9.6).

There are several ways of looking at the structure, but one of them is simply to look at the opposite curvatures of the paths as the top and bottom members of a beam with two hinges (figure 9.7) — a lens-shaped truss in the center and cantilevers at each abutment (figures 9.8 and 9.9). The public path is also what gives the structure its rigidity.

9.10
An acoustically efficient light structure for a bandshell in a park, Provence, France (2003), architect Explorations, engineer RFR Consulting Engineers.

BANDSHELL

The next example is a very small project, only a couple of hundred square meters: a bandshell for an open-air piano festival held in an old park in Provence.[3] Despite its modest size and apparently simple brief, a great deal of interaction between various factors occurs when seeking a single shape that would be equally satisfactory from the point of view of acoustics, structure and form: an acoustically efficient light garden structure (figure 9.10). The shell was first seen through an optical analogy, leading to an initial geometry. Modeled and explored through software that simulates the intelligibility of sound, the initial geometry yielded a first revision, itself the starting point for a set of analyses looking at the structural behavior of the shell. Increasingly detailed models helped understand how to adjust the geometry, global and structural, in a way that would remain compatible with the acoustic and structural behavior of the shell.

This type of exercise on such a low budget, small-scale structure has become possible only recently thanks to inexpensive, user-friendly, reliable software. Significant to what is happening in both architecture and engineering is that the tools change not only the quantity of things we can do but also the quality. Different types of investigations can happen much earlier in the design process. A great deal of interactivity, not previously possible, can now take place realistically. The change of paradigm has just begun.

9.11
The proposed extension of the train station in Strasbourg, France (2003–07), architect and engineer RFR Consulting Engineers.

STRASBOURG TRAIN STATION EXTENSION

The last project is an extension of the old train station in Strasbourg, again with the help of the SNCF's in-house architectural department and of Matthias Schuller's team in Stuttgart. Trains stop one level above the access level, while public transportation runs 15 m below ground. A link is required in front of the station, between the street level and the underground tram. However, the station is one of the urban features of the city and the new building extension should not hide it. Therefore the project seeks to maintain the historical presence of the train station, to link the various points of access to the various modes of transportation and to provide proper shelter in all possible climates.

The initial answer, a glass bubble along the south side of the station (figure 9.11), had to be considerably refined in view of the first thermal simulations. The skin of the building, a filter for solar energy and heat, dominates the design so much that the work focused first on resolving questions related to the behavior of the skin before tackling its supporting structure. Two questions arose. First, how can the geometry be resolved rationally to match objectives of space, shape and form? Generations by translation and rotation are being tested now. Second, what strategy would provide climatic comfort with a minimum expenditure of energy?

Various simulations have shown that only a combination of elements, adjusted to their orientation on the surface of the bubble, will let the skin shed

9.12
Section through the proposed extension of the Strasbourg train station.

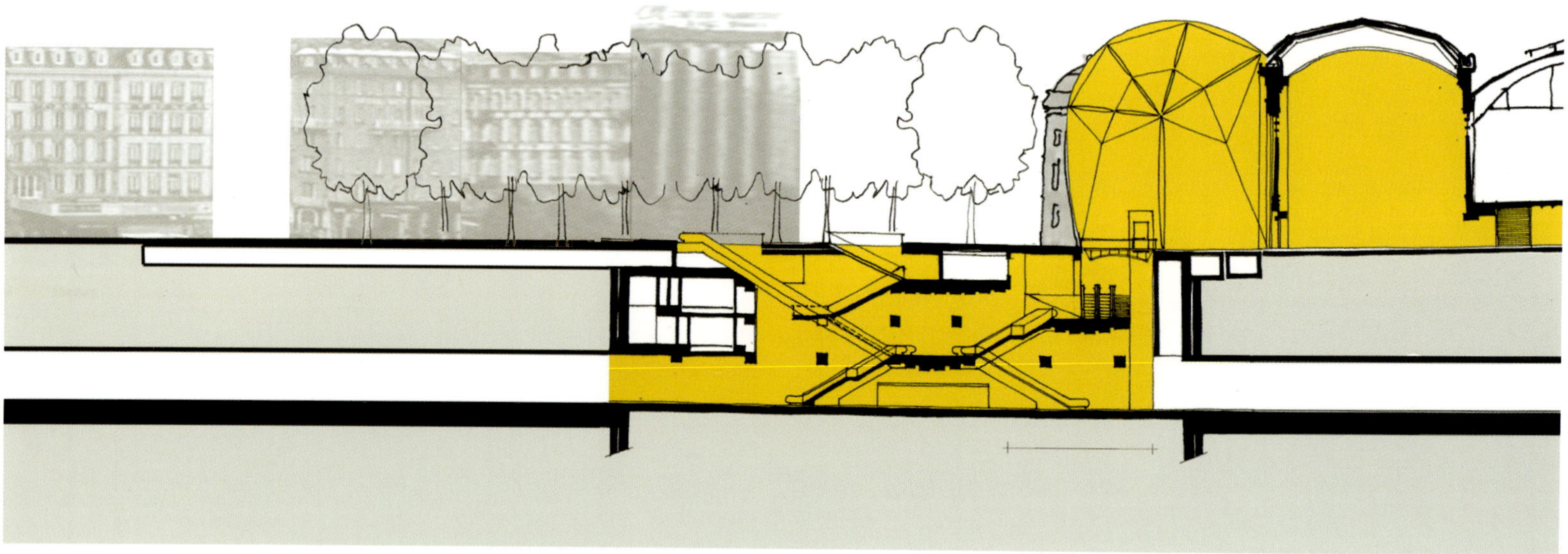

most of the incoming energy. The weather barrier is made of a glass combining a good heat mirror, ceramic frit screening, low-emission coatings, and something similar to a temporary double skin. A low-emission curtain drawn inside the glass bubble contains the heat at the surface of the glass and conducts it to the top of the building where it escapes through openings.

The most interesting discovery of this project was the unexpected opportunity offered by the tunnel dug under the station. Strasbourg is built along the River Rhine and the tunnel bathes in its waters, effectively turning it into a giant cooling system (figure 9.12). By carefully designing the building to help pull in more air than would come out naturally from the tunnel (and by adding a couple of simple, low-energy devices like a radiant slab, itself drawing from the water table), it is possible to make the space under this very large south facing glass wall comfortable in summer, almost exclusively through passive means. Now that the composition of the skin is known, we will design a structure that works with it. Again, the nature of the skin drives the organization of the structure and the nature of the skin is determined as much by optical criteria as by thermal ones.

CONCLUSIONS

This is not really a conclusion but a reaction to what was described earlier. Can we really separate architects from engineers by what can be described by numbers? Our culture, every single human culture in fact, can be defined by its technology. Why would one want to exclude technology from any discussion or be afraid of recognizing technology as part of our culture, to be understood, appreciated and judged? Ultimately it has to do with democracy — one cannot leave technology to the market. Architects have a responsibility to use technology well and not hide it — or hide from it — and bring it to the public.

NOTES

1 TGV – *Trains de Grand Vitesse* — France's high-speed train network.

2 The Bercy footbridge was designed by Henry Bardsley.

3 A competition won with Explorations, who are former students of ours.

10

NON-STANDARD STRUCTURAL DESIGN FOR NON-STANDARD ARCHITECTURE

HARALD KLOFT

10

NON-STANDARD STRUCTURAL DESIGN FOR NON-STANDARD ARCHITECTURE

HARALD KLOFT

10.1
The curved ribs of the *Palazetto dello Sport* in Rome, Italy (1956–57), architect and engineer Pier Luigi Nervi.

10.2
The Stadium for the 1972 Olympic Games, Munich, Germany (1968–72), architect Günter Behnisch, form-finding Frei Otto.

This chapter examines structural engineering in a performance-based architectural design environment, presented against the background of digital workflow processes of developing and producing the so-called "free-form" or "non-standard" architectures. Referring to the architecture of complex geometries as "freeform" is somewhat controversial because the term tends to simplify the conceptual and intellectual complexity of recently developed digitally-driven techniques of form generation. The curators of a recent exhibition[1] at Centre Pompidou in Paris instead used the term "non-standard architecture" to refer to the architectural objects expressing the new formal freedom. Whether "freeform" or "non-standard," a number of recently completed projects with complex geometries manifest a move away from traditional formal principles in architecture, often demanding novel approaches to structural design and engineering, in which performance-based design and digital workflow are closely interwoven.

What is "free" in a freeform? If one compares the biomorphic structures found in nature with the material realities of tectonic languages in building, an obvious difference is that there are no straight lines and rectangular geometries in nature. This is because the form generation processes in nature are based on the rules of evolution and are optimized for adaptability in the natural environment. In contrast, formal and tectonic languages in architecture are decided mostly by the capacities of the technical production, i.e. by available technologies in a given age. With industrialization, the technical possibilities of production increased substantially, making departures from basic geometries possible in architecture. The formal, geometric shifts often coincided with the development of new techniques and materials. This correlation is historically obvious; for example, in the 1950s, 1960s and early 1970s, developments in concrete and later plastics and textiles inspired architects and engineers to move away from Pythagorean geometries and to treat form in a less restrained geometric manner. Well-known examples are the filmy shell structures of Felix Candela, the structurally logical ornamentation of the *Palazetto dello Sport* in Rome built in 1957 by Pier Luigi Nervi[2] (figure 10.1), Eero Saarinen's concrete sculptural forms for the TWA Terminal in New York, completed in 1962, or Matti Suuronen's utopian plastic vision for "tomorrow's living" called the Futuro House. All these examples manifest the technical potential of this epoch and are expressions of the spirit of the time, like Frei Otto's lightweight tent- and cable-net structures which reached their zenith in the Munich Stadium for the Olympic Games in 1972 (figure 10.2).[3]

10.3
Endless House (1961), architect Frederick Kiesler.

10.4
Dynaform BMW Pavilion, 2001 International Motor Show, Frankfurt, Germany (2001), architects Bernhard Franken and ABB Architekten.

It is important to note that the formal "freedom" in that age (1950 to early 1970s) was limited because of the lack of high-performance hardware and software tools ("tools" in the sense of computer program components) in design development and manufacturing. Forms were often found in experiments with scaled physical models, as manifest in the work of Frei Otto or Heinz Isler. The forms they designed through a model-based form finding process were structurally optimized by following the rules of physics.[4] They stand in sharp contrast to the forms designed by form generation processes inspired by non-technical issues, for example in the work of Frederick Kiesler (who began his architectural formal research before World War II) and later on in the utopian ideas of Archigram and others.

Finding a structurally optimized and geometrically clearly defined form was a necessary condition for building, i.e. for material realization, in the pre-digital era. Frederick Kiesler, however, was not interested in defining forms in a geometrically exact manner that follows physical logic. For his design of "The Endless House" (figure 10.3), Kiesler made numerous freehand sketches to visualize his ideas about the form. His naturalistic design was celebrated as the "biomorphic answer and antithesis of the cubistic architecture of modernists."[5] For Kiesler, form does not follow function: form follows vision and vision follows reality. To communicate the spatial complexity of his ideas, he would create physical models just as a sculptor would model an art piece. Unlike the projects by Frei Otto and Heinz Isler, the form of Kiesler's "Endless House" was not inspired by structural optimization but by careful proportioning driven by the scale of human beings in the natural environment.[6]

ARCHITECTURAL FORM GENERATION

Recent years have brought a renaissance of freeforms — designers today are formally less constrained as a result of advances in computer technology. In Kiesler's form-generation process, the sketches and physical models were not linked — they had the same origin but the geometries were not exactly the same. Today, a digital model that links design and manufacturing processes is a basic condition for realizing freeform architectures. How the digital models are generated depends on the design process, which can vary distinctly from one designer to another or even from project to project. In addition, each geometrically complex project presents different challenges for structural design and can result in different digital working processes, as each of the four projects[7] described below demonstrates.

10.5
Dynaform: the force-field setup for the form generation process.

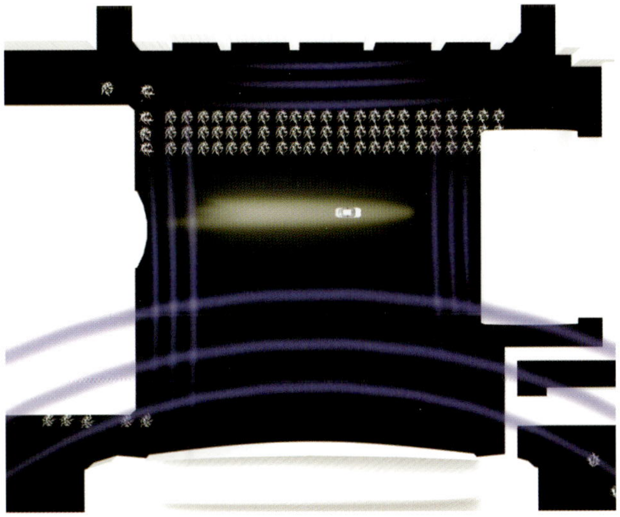

10.7
Dynaform: the "master geometry."

10.6
Dynaform: sampling of the form generation process.

Dynaform

Dynaform is the name given to the BMW pavilion project designed by Bernhard Franken (in association with ABB Architekten) for the 2001 International Motor Show in Frankfurt (figure 10.4). Bernhard Franken and his team use computational tools to generate architectural form; there is no preconceived formal idea at the beginning of a project. The "parametric design process"[8] of the Dynaform started with a briefing by the client about the communication message for the exhibition; BMW's primary interest was the presentation of the new "7" series. Franken translated program- and site-specific parameters into virtual forces using Maya animation software.[9] He set up a three-dimensional (3D) matrix that was initially shaped according to the virtual forces of a driving car (the influence of the program). Adjacent buildings on the site, such as Nicholas Grimshaw's Exhibition Hall, had additional impact on the shape through a series of specially designed force fields (figure 10.5). Through a time-based (4D) modeling process, the initial shape was deformed and altered by the software, until an architectural form was found by sampling the generative process (figure 10.6). The approximate shape of the found form was corrected for geometrical errors and established the 3D "master geometry" of the project (figure 10.7). That "master geometry" provided the dimensional reference for all project participants during design development and construction.

10.8
M.art.A. Museum, Herford, Germany (2000–2004), architect Frank O. Gehry and Associates.

10.9
M.art.A. Museum: 3D digital model of the building's geometry.

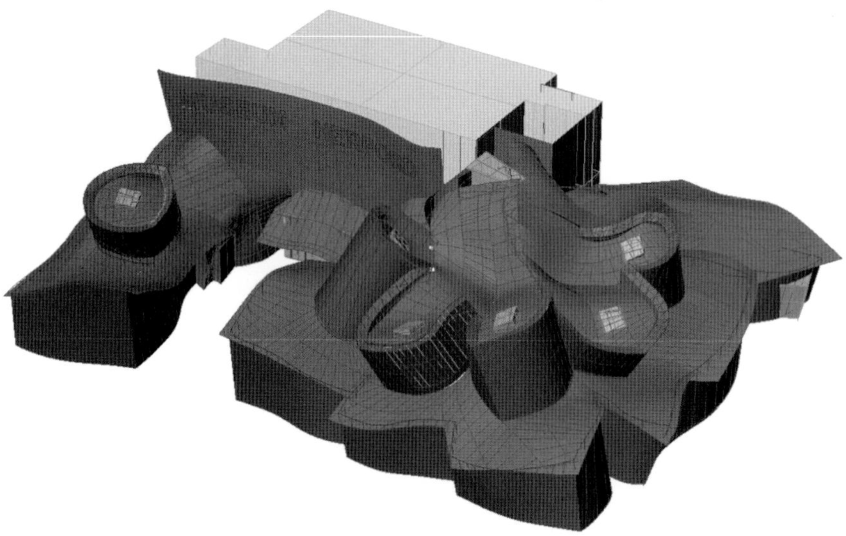

M.art.A. Museum

Frank O. Gehry's M.art.A. Museum (Museum for Furniture, Culture and Fine Arts) in Herford, Germany (2000–04; figure 10.8), illustrates a different approach. Here the design process did not start in a virtual design environment; instead, the architects manually built a series of physical models, many of which were three-dimensionally digitized in the computer-aided design (CAD) environment of CATIA[10] to correct and check the shape with respect to the program and the site. Gehry generally concentrates on the effects of the exterior surface and the interior spaces, and relies largely on physical models to verify that the original design intent is met. The CAD-corrected data (figure 10.9) enabled the building of more accurate physical models; partial and more detailed models are also often produced to explore the implementation of the overall shaping strategy in projects. Unlike Franken, Gehry did not define a 3D "master geometry" in this project as a dimensional reference before starting the design and structural development. Gehry's form-finding is an iterative process in which form changes are programmatically driven.

Kunsthaus Graz

Peter Cook's and Colin Fournier's design for the Kunsthaus Graz, Austria (2000–03; figure 10.10), illustrates the third approach. Here, the conceptual design phase (during the competition) did not rely heavily on computers and the physical model that represented the final shape was handmade (figure 10.11).[11] The first impulse was to perform a 3D scan of the competition model in order to produce an initial 3D digital model. But after considering how to optimize the form in terms of structure and materials, the architects and engineers jointly decided to generate a new 3D digital model from scratch using

10.10
Kunsthaus Graz, Austria (2003), architects Peter Cook and Colin Fournier (spacelab.uk).

10.11
Kunsthaus Graz: the handmade competition model.

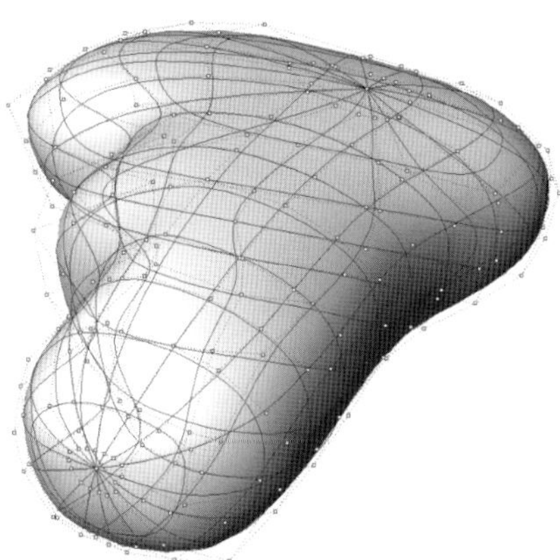

10.12
Kunsthaus Graz: the digital design model.

Rhinoceros 3D modeling software. While attempting to closely follow the shape proposed initially, the digital 3D model was built independently, i.e. it did not contain digitized data taken directly from the physical models, which is a process typically used in Gehry's office, as previously discussed. The digital design model (figure 10.12) was shaped to capture the design intent of the original scheme and allowed us to optimize the form during the design process with regards to the structural behavior, such as its geometrical stiffness, and to address some of the manufacturing issues. This is an important difference from the previously described form-generation processes used by Bernhard Franken, where the digitally-generated form was defined as a "master geometry," meaning that later in the structural design it was not possible to optimize the form geometrically, i.e. by changing the shape.

STRUCTURAL DESIGN

Structural design means moving from geometrical form to a structural system, i.e. connecting the formal architectural idea with the rules of the stress flow. What then are the implications of the different design approaches, discussed so far, for structural design? In all cases, it is important to note that the analytical software tools used by the engineers and the 3D design environments used by architects do not normally share a common digital database structure. Thus, it is in the interest of the engineers to develop data "post-processing methods" that automate and accelerate the geometrical data transfer.[12] Inputting geometry manually is impractical and time-consuming for complexly shaped structures. Being able to import data files accurately and quickly enables engineers to directly apply finite element and spatial vector programs to problems that need to be solved during the early design phases.

In Franken's Dynaform, a single layer skin defined the generated form — the "master geometry." The skin's material and thickness, however, were not specified. Since the "master geometry" was fixed, we could not optimize the structure geometrically through modifications of the overall shape. High local forces or bending moments had to be accepted because structural optimization of the overall shape would have called the underlying design approach into question. Instead, the aim was to support the intention of architecture to bring forces into a dynamic balance and to express the idea that the form is only a frozen instance from a long series of possible geometric constructs. The result was that the structural system had not been optimized in Dynaform.

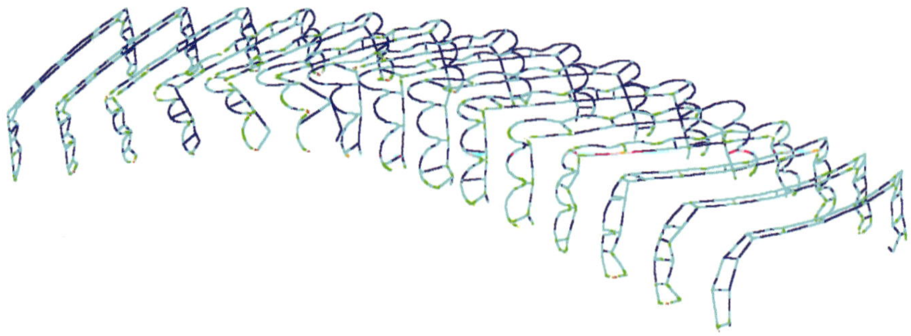

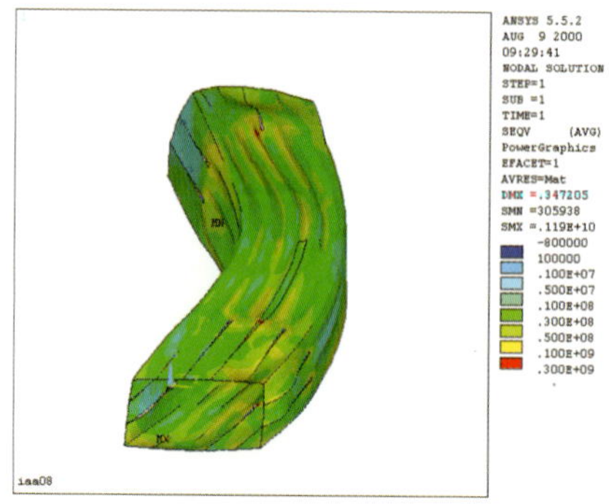

10.14
Dynaform: the finite element analysis of the master geometry in the pre-design phase.

10.13
Dynaform: the finite element analysis of the stresses in curvilinear structural members.

As there was little prior knowledge of structural performance of freeforms, we focused on two options to materialize the form and develop the structure: we could either design a system of linear or curvilinear structural members (figure 10.13) that would have supported a secondary and non-structural skin (figure 10.14), or the skin itself would have been conceived as the primary load-bearing system and would have become "skin deep"[13] — a surface-structure with shell-like behavior. The decision as to which option to adopt was driven by a number of studies of the different approaches; the final structural design was developed in close collaboration with the architects. The extremely short construction period was one of the main arguments to separate the structure and the skin by designing a primary load-bearing system of welded steel frames and a secondary pre-tensioned membrane layer.

In Gehry's project, we were given the digital geometric model of the outer and inner surfaces (figure 10.15); the interstitial, "in-between" space had to accommodate the structural and all necessary mechanical systems. The load-bearing structure — often a series of steel frames — was hidden from the view and consequently was almost irrelevant architecturally. In this "undercover" role, the main part of the structural engineering was geometric optimization, not with the aim of optimizing the geometric stiffness of the shape but to design a layout of structural members in the interstitial space and to optimize their arrangement so that they would work structurally (figure 10.16). The internal and external surfaces themselves acted as enclosures without any primary load-bearing function; their geometry established boundary parameters that generally could not be changed. As in Franken's project, the surfaces functioned as "master geometries" but these gave little opportunity for optimization of the structural behavior. Whereas the layering, i.e. the arrangement of the different functional layers (structure as an inside or outside layer, or in-between) and therefore the perception of structure, were open questions in the Dynaform project, in the M.art.A. project the structural elements functioned only in one dimension — as a technical necessity in-between.

10.15
M.art.A Museum: the architectural (non-structural) skin.

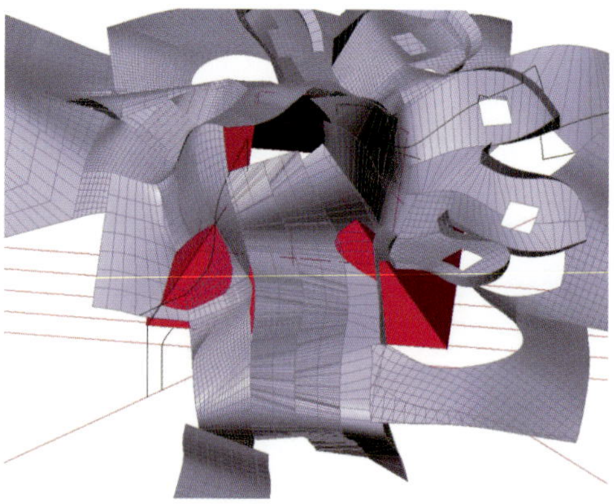

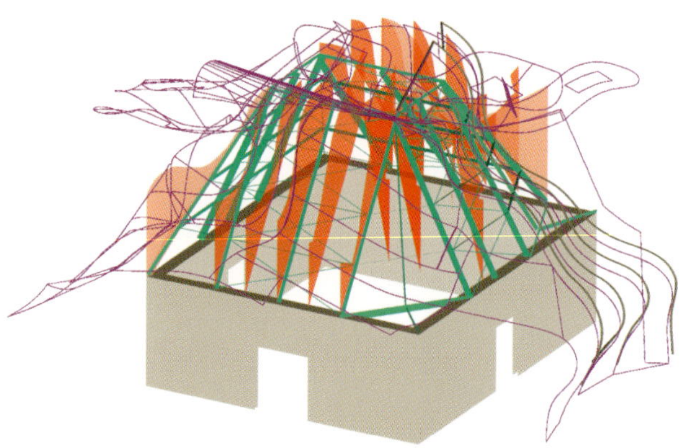

10.16
M.art.A Museum: the structural system was designed in the interstitial space between the given exterior and interior surfaces.

10.17
Kunsthaus Graz: the finite element analysis of the complexly shaped skin.

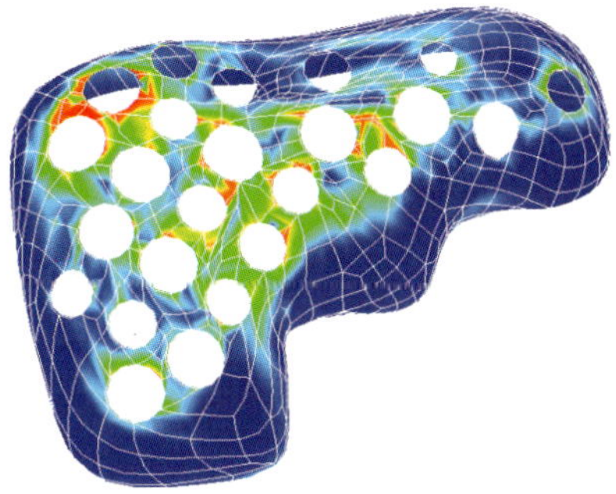

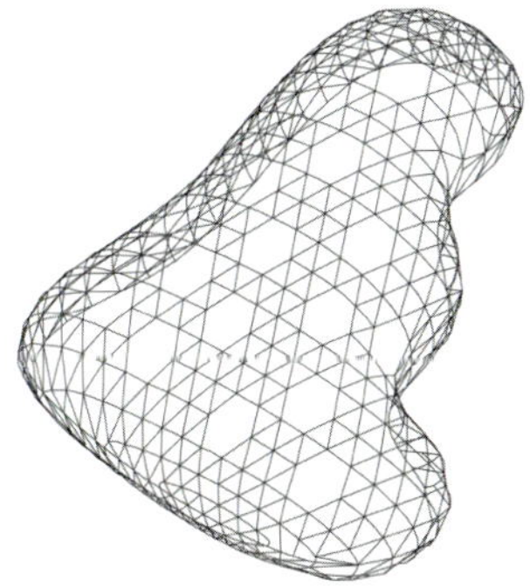

10.18
Kunsthaus Graz: triangulated structural engineering pattern.

Peter Cook and Colin Fournier, unlike Gehry, created the digital model of the geometry for the Kunsthaus Graz in the design development phase by translating the conceptual competition scheme into a digital environment. While the objective was to remain close to the original shape, Cook and Fournier did not pursue this dogmatically. Also, the structural behavior was allowed to have an influence on the final geometry.

For the complex roof of the Kunsthaus Graz, the principle task was to design a system of tubular steel members that would support an outer layer of acrylic glass panels with complex shapes (figure 10.17). The external skin was designed as a series of discrete layers, each responding to a specific set of functional requirements. The structural layer, in response to the need for a stiff and optimized structural system, consists of members arranged in a triangulated pattern that was designed as a hybrid structural system combining behavior of shell structures and bending systems (figure 10.18).

The latest project of Bernhard Franken — "Klöpp" — a ministry complex of three buildings in Reykjavik, Iceland (2002–; figure 10.19), deals with the structural design in a very different manner than discussed previously. It was the first project in which Franken did not strictly define a "master geometry;" the geometry, especially in the façades, was subject to change as the project developed. The generation of form was done digitally as in his earlier projects, using Maya animation software. The conceptual origin, as inscribed into the parameters of the initial form generation process, was found in the fissured surfaces of Iceland's topography and in the stone monument near the site, inside which, in accordance with the Icelandic legends, elves should live. The form generation was highly iterative: the initial form, generated by sampling time-based processes modeled in Maya, was structurally analyzed using finite elements software (figure 10.20a). After interpreting the results, we optimized the form geometrically in collaboration with the architects, while adhering to the initial design intentions (figure 10.20b).

10.19
"Klöpp" ministry complex, Reykjavik, Iceland (2002–), architect Franken Architekten.

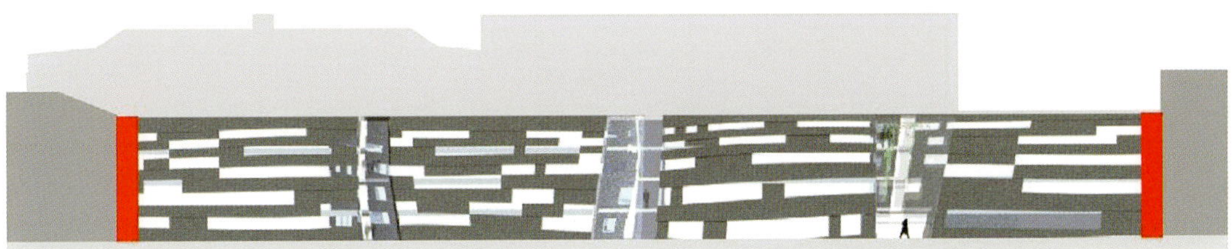

10.20a
Klöpp: finite element analysis of the outer façade.

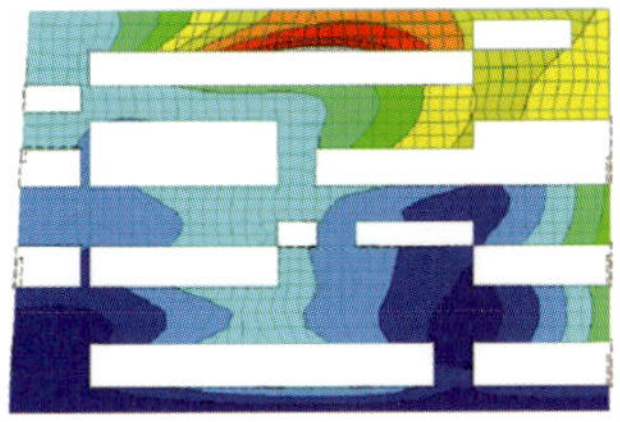

10.20b
Klöpp: structural optimization of the bracing system in the outer façade walls.

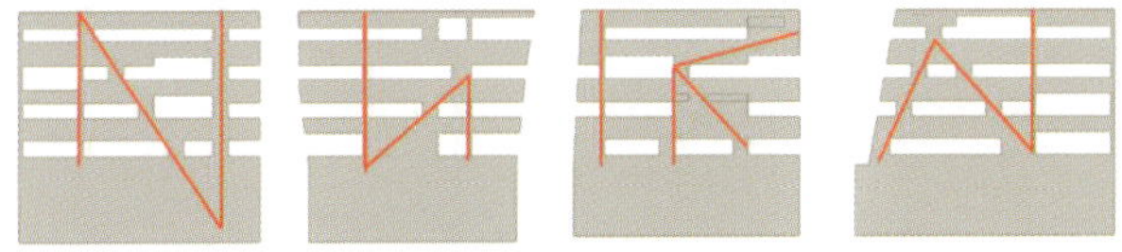

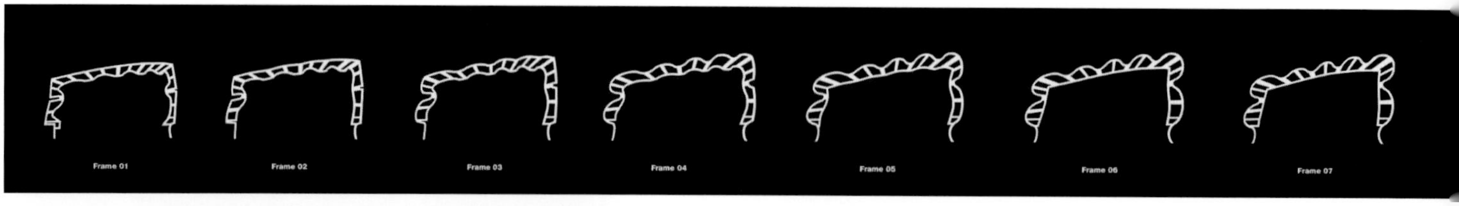

10.21
Dynaform: cross-section outlines as "derivatives of the master geometry."

10.22
Dynaform: the structural "Dynaframes."

10.23
Dynaform: the full-scale mockup of the "Dynaframes."

10.24
Dynaform: the full-size mockup of a section of the monoaxial pre-tensioned PVC membrane.

MANUFACTURING PROCESS

A lack of suitable materials and production techniques that are usable in building are problems that anyone seeking to design and build freeform surfaces is faced with at some point. Building regulations and standards that do not cover new procedures and materials, and potentially high construction costs, add to the difficulty of constructing a complex shape. Since buildings are normally prototypes, we cannot apply industrial processes that may be well suited for the production of parts with complex shapes but require large quantities in order to be economical.[14]

In the Dynaform project,[15] after much discussion and the evaluation of different options, the architects and engineers decided jointly to separate the primary load-bearing structure from the structural secondary skin and to design a series of primary steel frames. Franken's team generated fifteen cross-sections as "derivatives of the master geometry," each at a different angle (i.e. they were non-parallel) and each resulting in a unique shape (figure 10.21). We then inscribed structural frames — the so-called "Dynaframes" — into those sections. The outer outline of the Dynaframes is precisely offset from the "master geometry" surface, while the inner outline represents a reversal of the same master shape (figure 10.22). Both outlines are connected at regular intervals with welded plates in order to work structurally as a Vierendeel system. Those plates all point to the origin of form generation — the virtual curvilinear path generated by a car driving through the space (figure 10.5). The design of the rather strange looking Dynaframes thus originates in the general form-finding principles of the scheme — it is not primarily driven by a structural logic.

The steel members for the frames were cut from flat steel plates, then bent into shape and manually welded together. The contractor was faced with the challenge of having to maintain tight tolerances while translating the digital 3D data into a built shape. A full-size mock-up of several structural frames was completed three months before the final erection (figure 10.23); it enabled us to study the required assembly time and procedures as well as identify and resolve problems with the connections between certain components.

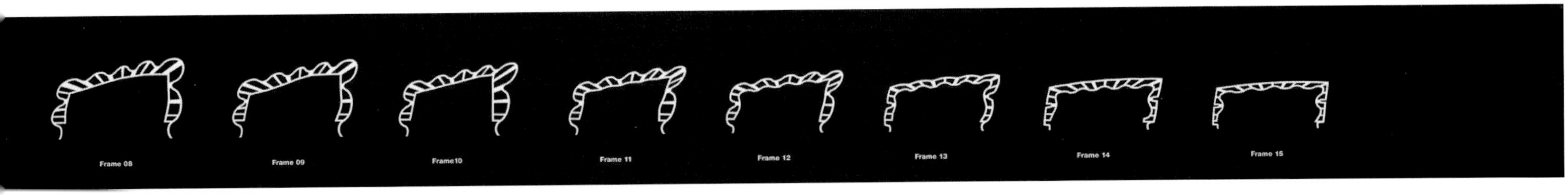

10.25a
Dynaform: assembly of the membrane in the most difficult area — the transition from the curvilinear to rectangular surface.

10.25b
The Dynaform with open joints.

10.25c
Dynaform: the skin fits tightly.

The exterior of the pavilion is meant to be the purest possible representation of the "master geometry." A major concern was to find a way of constructing a skin that would generate a smooth surface over the complex shape. After researching different material options, a monoaxial pre-tensioned PVC membrane was selected.[16] By modeling the skin as a ruled surface, we avoided any external folds, and the membrane could be pre-stressed between the structural frames. To prove that this new envelope concept would work, a full-size mockup was constructed and subsequently approved by the design team and the client (figure 10.24).

The membrane layer was assembled by specialists with mountain-climbing skills (figure 10.25). The assembly started two months before the opening of the motor show. Each span between adjacent steel frames was covered with one membrane segment. The joints between membranes were sealed with an apron fabric (figure 10.26) — a compromise as far as the overall appearance was concerned, but unavoidable with respect to the tight construction schedule and the need for a watertight envelope.

10.26
Dynaform: the joints between membranes were sealed with an apron fabric.

10.27
Dynaform: the seamless materiality with complex geometry.

The finished project looked effortless, with materials and shapes appearing to connect almost naturally (figure 10.27). Designing and building a large freeform structure, such as the Dynaform, requires more energy, time and creativity than would be necessary for a conventional structure. Deadlines, the budget and the design intentions could not be compromised. The Dynaform pavilion remained in place for the duration of the motor show; it was then disassembled and is stored for future use.

CONCLUSIONS

The translation of freeform, non-standard architectural designs into built structures requires the development of new modes of thinking from all project participants. It is essential that architects and engineers collaborate from the very beginning of a project. In the case of "freeform" architecture, an important aspect of this collaboration is that the structural engineer has to "speak the language" of the architect and fully support the particular design approach. Understanding individual design values is essential for a productive collaborative dialogue between engineers and architects.

10.28
Dynaform: the full, detailed 3D model of the building data.

The shown projects mark the beginning of new opportunities in building. Boldly curved shapes, like the Dynaform and the Kunsthaus Graz project, were regarded as unrealizable a few years ago. Both projects are radical in the way all participants — clients, planning parties and contracted firms — believed in the idea of realizing a visionary architecture and took the risk of failure. The planning and manufacturing processes were characterized by the necessity of developing new solutions from scratch. In this context, it was extremely helpful that Franken's parameter-driven design method in the Dynaform project did not allow for geometrical alterations to the architectural model. Only with these experiences is it possible to take on the challenges of the "Klöpp" project, where the geometry, especially in the façades, is subject to change for structural optimization. A collaborative process based on the mutual processing of parametric data, as was the case in the "Klopp" project, is the most promising for the future, since it allows for the geometric optimization of form in terms of its structural behavior.

In my experience, there are several "rules" for engineers and architects interested in designing and building the "non-standard architectures." First, full-scale mockups of the geometrically most complex and challenging areas are absolutely essential. Problems that cannot be solved in the production of the mockup will not be solved on the construction site either. Second, the digital model of the geometry for use in manufacturing and construction should be fully parametric in order to facilitate the quick generation of the necessary data sets. In the Dynaform project, for example, the specialists for structural membranes developed a method for generating a ruled surface from the given geometry by splitting the membrane into pieces which seamlessly connect the cross-sectional frames in areas of similar curvatures; that is the principal reason why we could use the uni-axial tensioned membrane without getting folds on the surface. Third, the control of layering during the manufacturing process should not be left to the construction company. This is an important new task in the digital workflow and should be done by the design team, either by architects or engineers. As a task, the layer control is analogical to the role of a compiler in the computer — it produces the detailed "machine" code, i.e. the data necessary for execution. In the Dynaform project, the 3D model of the geometry produced in the architectural form generation process is one-layered — it consists of surfaces without material specificity, without thickness. At the end of the design development, a multilayered 3D model is produced, which includes all the information necessary for manufacturing (figure 10.28). During manufacturing, however, that 3D model undergoes numerous changes depending on the tolerances and specific demands of production. As in compilers from the computer world, the task of surface control is to provide each manufacturer with a specific layer of information that is needed.

In conclusion, the emerging digital design and production environment, combined with new materials and modern technologies, offers unprecedented possibilities for architecture that strives to perform fully in a conceptual, formal, technical, financial and material sense. In such a context, structural engineers can exploit fully the creative potential of the discipline, not only by supporting the architects with the calculations, but by embracing their ideas and realizing the digital form generation possibilities structurally and materially. The freeform, non-standard and performative architectures offer a promise of new collaborative design synergies for architects and engineers.

NOTES

1 "Architectures Non Standard," December 10, 2003 – March 1, 2004, Centre Pompidou, Paris, France (http://www.centrepompidou.fr).

2 A common question is whether the chosen structure is the most logical one. Klaus Bollinger and I agreed when we visited the *Palazetto dello Sport* in 2003 that the structural system could have been realized in a variety of more logical ways; much of the necessary, basic structure could have been simpler. Nervi was obviously interested in expressing the aesthetics of structure in his buildings. He did not hesitate to explore the decorative, the ornamental, which was a taboo for the architects at that time. The ornamentation was achieved with curved ribs, logically oriented in the direction of the stress flow. One could imagine that the architects envied the engineers for this chance to deploy the decoration for purely structurally reasons (i.e. "form follows function").

3 Günter Behnisch won the competition; the final design was developed in collaboration with Frei Otto.

4 Model-based form-finding methods were used in earlier times too. Famous are Antoni Gaudí's physical models for the church of *Sagrada Familia* in Barcelona. Gaudí was obsessed by finding the structural and material given limits, which is why he investigated every detail in scale models.

5 Harald Krejci in Frederick Kiesler, *Endless House 1947–1961*, Frankfurt, Vienna: Hatje Cantz Publishers, 2003, p. 12.

6 At a Kiesler Symposium at the MMK (Museum of Modern Arts) in Frankfurt, Greg Lynn asserted that Kiesler did not proportion his drawings and models. That is quite true in the sense of a geometrical and architectural proportion theory, such as the harmonic proportion theory of Andrea Palladio. But I think Kiesler did proportion consciously his design for "The Endless House." This difference in understanding Kiesler's use of proportion can be illustrated with an analogous difference between the eastern idea of music and the western harmonic theory. Whereas in western culture composers think within a defined geometrical system of standardized pitches, in eastern cultures the atmosphere of the single sound counts, and the rhythm and time is realized as part of the nature. John Cage, for example, has integrated these natural aspects of eastern music in his compositions.

7 I worked as a project leader on the first three projects (Dynaform, M.art.A. Museum and Kunsthaus) in the office of Bollinger + Grohmann; the structural design for the fourth project (Klopp) was developed by osd — office for structural design (http://www.o-s-d.com) — which I founded in 2002 together with my partners Klaus Fäth, Sigurdur Gunnarsson and Viktor Wilhelm.

8 As referred to by Franken.

9 Maya is developed and marketed by Alias Wavefront (http://www.aliaswavefront.com).

10 CATIA is a 3D software tool for virtual design, simulation and analysis in industrial product processes. It is developed by the French company Dassault Systems. The breakthrough of CATIA goes back to the development of the 777 airplaine by the Boeing Company in the late 1980s. This airplaine was completly designed in the CAD environment of CATIA.

11 The form-generation process used in the Kunsthaus is, of all the projects discussed, the most comparable to Frederick Kiesler's. Peter Cook and Colin Fournier sculpted the shape of what was later called "Friendly Alien," as an artist would, in one piece, in difference to Gehry, who assembles his geometries by composition of several different components. The proportions of the form were inspired by the site, embedded in the old part of Graz with its typical, red-colored roof landscape.

12 In a digital architectural model, the geometrical data are based on surfaces or volumes. Analytical software tools — finite elements and spatial vector framework programs — presuppose definitions of mathematical and mechanical conditions: first, an approximation of the surface or volume by a mostly triangulated pattern and, second, definition of the connection conditions between the pattern elements. For example, points of intersection, which are geometrically defined in an architectural model, have to be defined in analytical software with regards to their mechanical movabilities — degrees of freedom. Depending on the mechanical conditions, one can more or less force a structural system into a given geometrical form.

13 "Skin deep" was articulated as such by Johan Bettum in a lecture at the Städelschule in Frankfurt.

14 This is a major difference to design processes of industrial products. While high-tech products are mass-produced, nearly every building is unique, often realized under extreme pressure in a single manufacturing process. As a consequence, the application of innovative materials and techniques in building construction progresses at a much slower place. However, it is only by the integration of new technologies in building that we are able to realize complex shapes. For example, the Experience Music Project (EMP) in Seattle (1998–2002), designed by Frank O. Gehry and Associates, could not have been built without the use of Global Positioning System (GPS) technology. The outer surface of the EMP was defined by the digital architectural model. The distance between the outside layer of metal sheets and the waterproof layer of sprayed-on concrete underneath was measured by GPS devices; the measurements provided the data for producing the fixings.

15 The description of the manufacturing process of the Dynaform project is a good example of the expense of manufacturing complex shapes; in my experience there are, in comparison to design phases, no fundamental differences in the manufacturing process between the described projects.

16 This method was developed by Viktor Wilhelm, who is a partner in osd.

11

COMMUNICATIVE DISPLAY SKIN FOR BUILDINGS: BIX AT THE KUNSTHAUS GRAZ

JAN EDLER

11

COMMUNICATIVE DISPLAY SKIN FOR BUILDINGS: BIX AT THE KUNSTHAUS GRAZ

JAN EDLER

11.1
The BIX communicative display skin, designed by realities:united, integrated into the main eastern facade of the Kunsthaus Graz, Austria (2003), architects Peter Cook and Colin Fournier (spacelab.uk).

Most buildings that surround us are static — to accommodate change, they are either reconstructed or demolished. Generations of architects and engineers have dreamt about buildings and structures that can literally *perform* in order to adapt quickly to varying needs or circumstances by changing the physical shape, spatial and functional configuration, levels of natural and artificial light, overall aesthetic appearance, etc. Yet, until very recently, most of these visions have hardly been realized, primarily due to financial and technological limitations. Those that had been built were usually limited to small-scale or highly visible, high-budget projects.

Whereas the development of a physically changeable (robotic) architecture appears still to be many years away, there is a promise of significant progress in digital display technologies, which allow patterns, images, text, etc., to be mapped onto a building's surfaces, thus changing its appearance. The decreasing costs of the "big screens" have already caused an increase in the application of electronic advertising boards on building façades. A common practice in high-density urban areas is to turn the façade into a billboard to display mostly foreign (i.e. not building- or area-related) messages of globally active industrial players.

11.2
Kunsthaus Graz, as seen from the Schlossberg fortress above the city.

SKINS THAT PERFORM: AN ARCHITECTURAL ISSUE

As the technological progress continues — and the display technology becomes cheaper — the question of why and how to apply this type of technology to buildings will fall back to the realm of architecture, because the definition of the relation between a building and its outer surface is one of the oldest and most essential aspects that define architectural thinking. This will not be different when the outer appearance starts to change more rapidly (than once every twenty years). It is a change from the static "monument" to a performing "actor" that still remains in the realm of architectural consideration at first. "Architecture," however, is not ready for the task. Today we are missing clear and convincing architectural concepts for the design of media façades, in addition to the content that they should show.

As soon as a building façade becomes a "media façade," the ownership changes as well; this is appropriate if the façade is used for direct branding by the company that occupies the building. But a majority of such installations broadcast global advertising messages, thereby denying any form of relation between the specific building and its outer appearance. The surface of the building becomes separated and alienated from its inner programmatic structure, a trend that is also evident in more ambitious architectural or artistic experiments.

Labeled "interactive" (which is meant to stand for participation and democracy), the façade is turned, for instance, into a reflective device for any passerby who, willingly or not, controls the appearance of the building, facilitated by some sensor-driven computer software. Some of that "interactivity" is appealing, but is actually completely arbitrary in most cases. The current temperature, time, or some other banal piece of information, reoccurs time after time, and is apparently judged to be important enough to be broadcast at large with considerable technical effort.

In other words, the widespread ambition to equip buildings with media surfaces, and the available concepts for content, do not match. It is important to be completely clear about the problem — it is not the financial pressure of some "evil" industry intending to conquer architectural surfaces to turn them into advertising billboards. The problem is a lack of cultural or aesthetic concepts that are strong enough to be perceived as beneficial or even necessary for the appearance of buildings (and, by extension, cities). We need valid dynamic aesthetic concepts — *choreographies* — as a continuation of the architectural culture that took centuries to develop!

11.3
Kunsthaus Graz: detail of the building skin constructed out of individually shaped blue acrylic glass panels.

11.4
Early (2001) conceptual rendering of the BIX façade (Kunsthaus Graz, 2003), designers realities:united.

THE BIX PROJECT

BIX[1] is a light and media installation for the Kunsthaus Graz,[2] Austria (2003; figure 11.1). It transforms the main eastern façade of the building into an alterable, performative membrane to transmit internal processes of the art institution to the public. It is an attempt to create an experimental laboratory for the development and deployment of a unique urban communication style — a "language" — which is synchronized with the architecture and its users on the one hand, but which is also deferring to its urban context on the other.

BIX has the potential to become a reference project in the discourse about architectural performativity and the so-called media façades. It is the first very large, permanent urban screen installation, run by a non-commercial entity and exclusively designed and dedicated to show artistic productions.[3] A contemporary art museum is especially adequate as a "host" entity for this project because, by its mandate, it is expected to experiment and map the potential of new communication tools, such as BIX.

Background

The city of Graz in Austria was the European Cultural Capital for 2003.[4] As part of that cultural project, the city commissioned several new buildings, including the new "Kunsthaus Graz" — a museum for international exhibitions of modern and contemporary art, which was inaugurated in September 2003. The building's spectacular design stems from the award-winning competition design by the 1960s Archigram[5] legend Peter Cook, his partner Colin Fournier, and their spacelab.uk team (figure 11.2).[6]

The irregularly shaped, biomorphic building structure floats as an independent body, balloon-like, above a glass foyer. The sleek, blue shimmering façade is the outstanding characteristic of this building, referred to as "Friendly Alien." It is constructed from more than 1,100 individually shaped, translucent acrylic glass panels, wrapping the whole volume of the building like a skin (figure 11.3).

The Berlin-based architectural studio realities:united took the building's skin a step further and turned it into a giant media screen called BIX.[7] The BIX concept was initiated and developed in the summer of 2001,[8] at a time when the overall planning had already reached a very advanced stage (figure 11.4). In addition to the technical complexity of the project and its advanced development

11.5
Exploded view of the BIX matrix as part of the eastern Kunsthaus Graz façade.

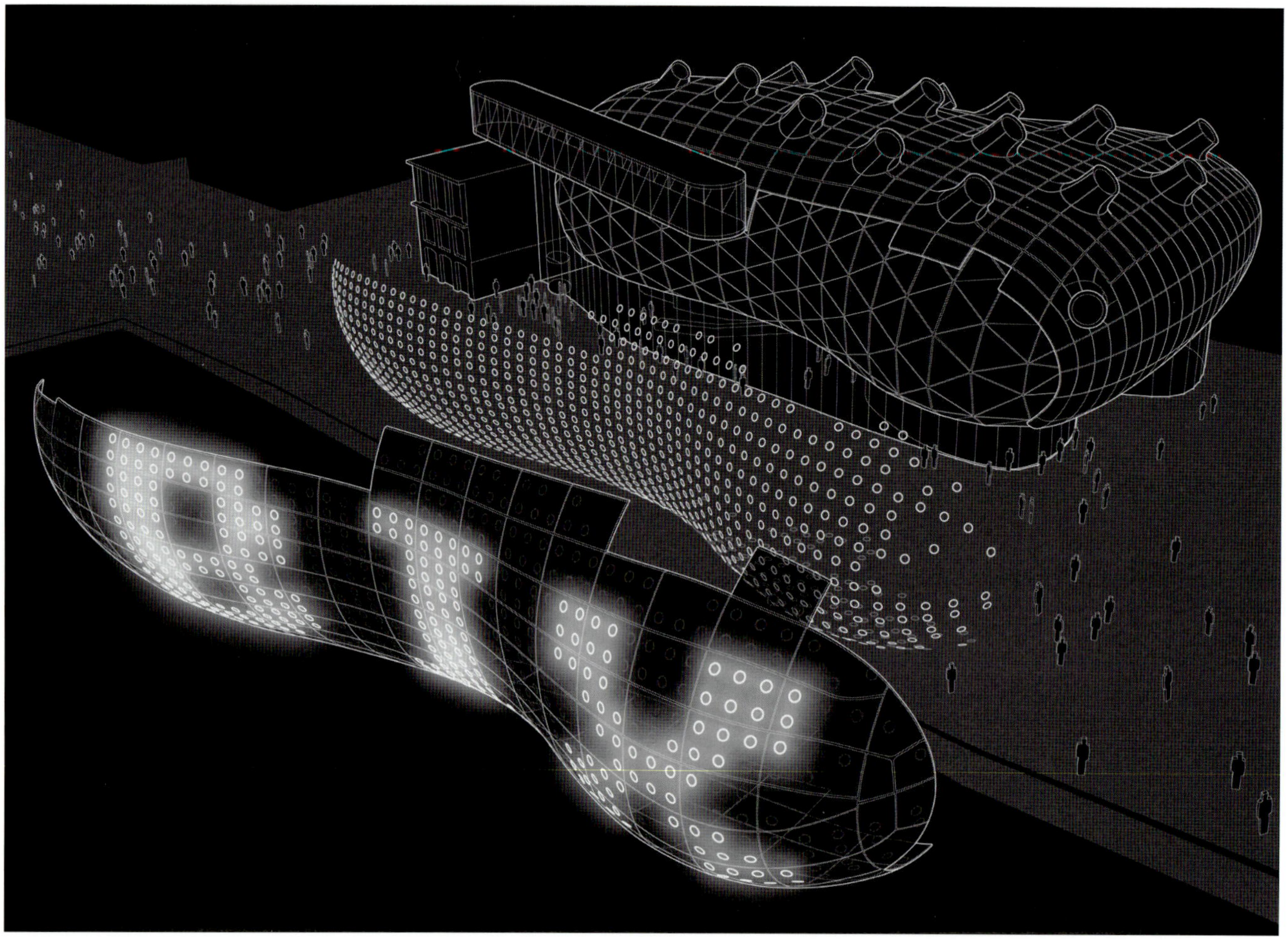

11.6
Installation plan for the BIX matrix.

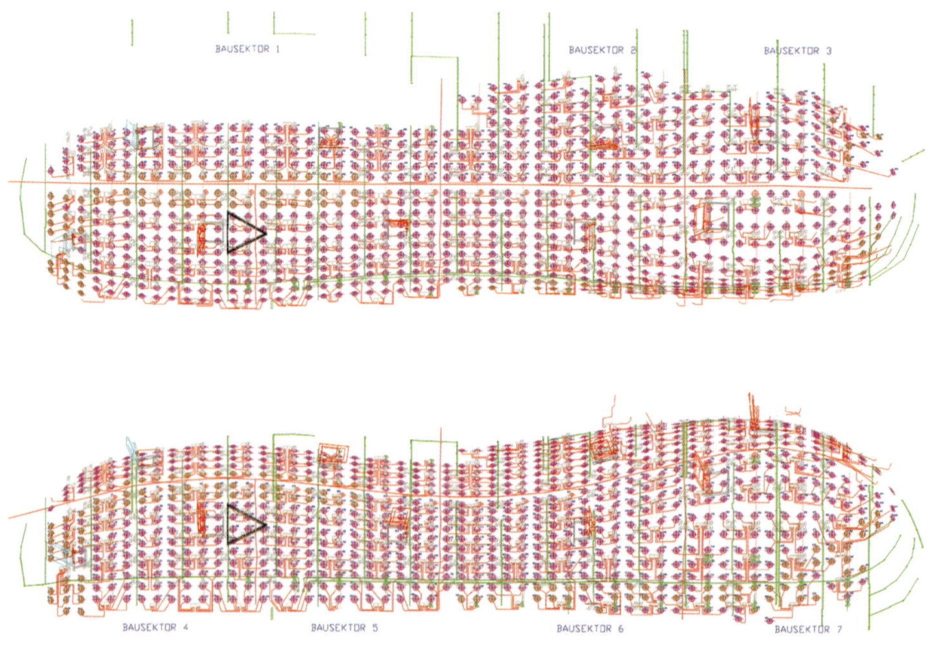

11.7
BIX: installation of the fluorescent lamps as independent modules prior to the mounting of the building's skin.

stage, integrating an architectural concept of foreign authorship into such an expressive building design was also a challenge. After all, BIX was a new element designed to dominate the building's riverside frontage, thereby radically redefining the architectural concept of the building's skin. At the end of 2002, the BIX project received approval by the client and the Kunsthaus Graz architects.

11.8
The BIX matrix has a low resolution of 930 pixels (lights).

11.9
Resolution of the BIX matrix (white rectangle) versus a typical TV screen (black rectangle).

11.10
The size of a BIX movie/animation — 930 pixels — as it typically appears on the computer desktop (scale 1:1).

Description

Beneath the acrylic surface facing the Mur river and the city center, realities:united deployed a matrix of 930 fluorescent light rings covering an area approximately 20 m high and 40 m long (figures 11.5 and 11.6). Each light ring acts as a pixel (picture element), whose brightness can be computer-controlled and infinitely varied at the rate of 18 frames per second. In this way, low-resolution light patterns can be generated over the entire façade and be visible from a considerable distance all over the city.

Each of the matrix's individual pixels is a conventional 40W fluorescent lamp with a diameter of 40 cm (figure 11.7, and see later figures 11.13 and 11.14). The decision to use this industrial module exemplifies the asymmetrical design character of the BIX concept. The design features of conventional large-screen displays were abandoned in order to obtain a number of substantial advantages in return.

The "resolution" of the matrix is extremely low. There are only 930 pixels — a mere 0.2% of the pixels found in a typical TV screen (figures 11.8–11.10). In addition, they are monochrome only. On the one hand, such a low image resolution imposes strong limitations; on the other, however, it enables both the modular structure and the large size of the installation to be highly integrated into the architecture. The BIX installation covers virtually the entire façade facing the riverside. Using conventional "big screen" display technology,[9] and with the same budget, the covered surface area would have been nearly a hundred times smaller.

11.11
BIX: a new standard in fusing architecture and media technology.

Enabling architecture

Sharing the same scale, the architecture and the media installation together generate new aesthetic results. It is not a separately mounted video wall but the Kunsthaus itself that radiates characters and images; the projection and the building achieve an extremely high level of integration as a single entity. The 930 lights of the BIX installation seem to be "tattooed" onto the skin of the building like individual spots of pigment (figure 11.11).

The light matrix does not constitute a rectangular field with straight sides, but rather an amorphous zone tailored to the complex shape of the building and gradually fading away towards the edges. Hidden behind the acrylic glass façade, only the active light rings are visible, while the rest of them remain invisible behind the skin so that the installation's edges are not always perceptible. Thus one has the impression that the blue bubble itself renders the light patterns from within. In the absence of a recognizable boundary, it looks as though the light patterns could dance freely on the outside skin of the building.

BIX, however, offers the Kunsthaus Graz significantly more than just a spectacular presentational touch because the installation also acts as an architectural "enabler,"[10] realizing the envisioned concept of the building's skin

11.12
By translating the original architectural concept of the building, BIX acts as an architectural "enabler."

(figure 11.12). In the winning competition entry in 2000, architects Cook and Fournier described the skin with the following properties:

> Much of it is opaque, but from time to time there are revealing slivers of transparency or hints of the presence of action within. Strange things appear and disappear within the skin: signs, announcements, short sequences of film or images: glimpsed for moments, only to fade away. For this sleek cocoon is a membrane that hints of new and creative activities within.[11]

The BIX project adopted these architectural pretenses at the time when it became clear that the initial concept of a mostly transparent skin would have to change for technical reasons. The installation preserves the original design intentions, even if in a mediated way. In this way, BIX was not only realizing the original architectural vision but the installation became an important argument in justifying the expensive acrylic façade panels to be mounted in front of the inner non-transparent, hermetically sealed off, bubble-like shell structure.[12]

11.13
BIX: computer rendering showing the fluorescent light tubes as a layer beneath the acrylic glass panels.

11.14
BIX: installation of the acrylic glass panels over the light tubes.

Technology

The choice to use "low tech" fluorescent light tubes (figures 11.7, 11.13 and 11.14) as the basic module for the display addresses the issue of *technological sustainability* over time. In comparison to architecture, new technologies for large screens age, i.e. become outdated, at a very fast rate. However, by using conventional, circular fluorescent lights for the BIX pixels, known since the 1960s as kitchen lamps that are almost a design classic today, the question of being technologically up-to-date does not arise. By using the fluorescent light rings, i.e. an "outdated" technology, the BIX display meets the architectural demand of constancy. This central attribute of the installation — technological sustainability — saves the operator constant upgrades and guarantees an operational balance between architecture and technology at comparatively low costs.

The BIX installation is driven by custom-developed specialized software tools, which are of the utmost significance for the efficiency and precision of the creative productions that are to be shown on the façade. There are two major software modules: the "BIX Director" and the "BIX Simulator."[13]

The BIX Director application allows the user to compose and schedule a program to be shown on the façade. The application's interface is similar to those of popular video-editing environments. Four different video tracks are available for arranging and mixing multiple "events"[14] on a 24-hour timeline. In this way, the Kunsthaus Graz can set up complex "shows" for particular days or for several weeks in a row.[15]

The second software module — the BIX Simulator — is even more crucial for artistic productions. It enables artists to examine the results in a real-time three-dimensional (3D) computer simulation of the Kunsthaus Graz in its historic context. By navigating through the city as if using a 3D "shooter" game, artists can ensure that their productions adapt to the large-scale complex geometry and the coarse resolution of the façade display (figure 11.15).

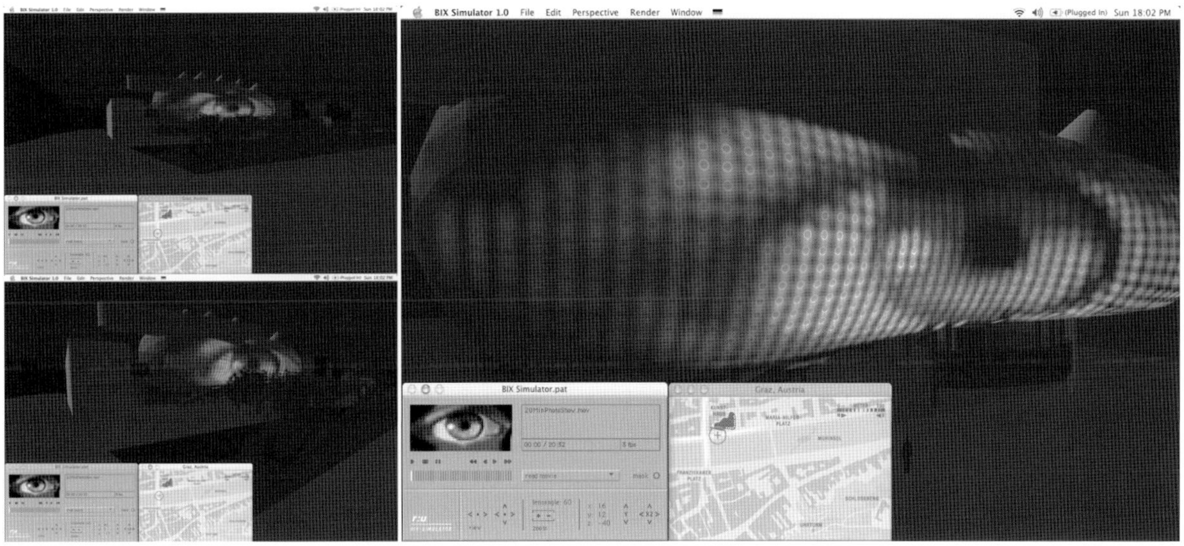

11.15
Screenshot of the
BIX Simulator
3D application.

BIX AS A COMMUNICATION LABORATORY

The BIX media installation and the Kunsthaus Graz's architecture share a strong symbiotic relationship. The façade as a display extends the communication range of the Kunsthaus Graz, complementing its programmatically formulated communicative purpose. In an abstract and mediated form, the media façade transmits the internal processes of the Kunsthaus Graz out into the public, creating a symbiosis of art, architecture and media. BIX, therefore, becomes an important identity — and an image-building factor — of the Kunsthaus Graz.

Peter Pakesch, Director of the Kunsthaus Graz, sees in BIX a new level of art mediation:

> I think that the architects and especially realities:united, the creators of BIX, have succeeded in presenting a different kind of transparency — it is not the superficial kind of transparency of a glass house which is useless for an art museum anyway, but more a transparency of information, the translucency of content, for which a lot is still to be developed and for which the architecture is a challenge.[16]

If a cultural institution like the Kunsthaus Graz is a tool for artistic articulation, the BIX installation multiplies its power by turning the Kunsthaus Graz into a "power tool," where the power is not defined in a physical sense but above all by a capacity to articulate and broadcast meaning. As the content producer, the Kunsthaus Graz has the chance, as well as the responsibility, to develop methods for a dynamic communication between the building and its surroundings, between content and outside perception. Hence, a unique form of communication, consisting of vocabulary, syntax and rhythm, needs to be developed.

At the same time, the communicating skin is a unique experimental working platform for art projects investigating forms of interaction between media and space (figure 11.16). With BIX, artists can explore alternative cultural and artistic modes of production, whose implementation on commercially used "propaganda" surfaces is widely excluded. The enormous size and the rough resolution of the installation in comparison to conventional display systems aim at the core aspects of artistic research: reduction and intensity are well-established strategies of contemporary art to advance towards the inner essentials. In this way, BIX not only extends the Kunsthaus Graz's communication range — both spatially as well as temporally — but the installation replenishes the overall program of the Kunsthaus Graz.

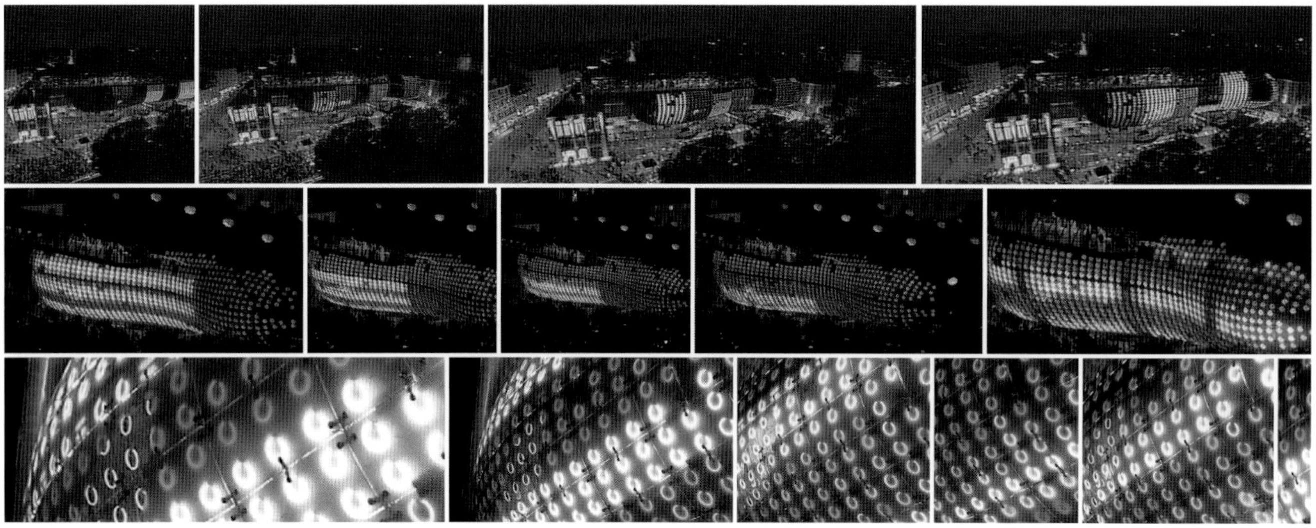

11.16
BIX as an artistic communication laboratory: video stills from the live audio visual performances by artists John de Kron (Berlin) and Carsten Nicolai (Berlin) for the inauguration of the BIX installation in September 2003.

NOTES

1 http://www.bix.at
2 http://www.kunsthausgraz.at
3 Obviously BIX is not the first non-commercial media façade. But most façade projects we know of were either temporary installations or are commercial advertising boards. An overview of interesting media façade installations and related projects can be found on the BIX project website at http://www.bix.at
4 http://www.graz03.at/
5 http://www.archigram.net/
6 The competition entry was designed by Peter Cook and Colin Fournier with Niels Jonkhans, Mathias Osterhage, Marcos Cruz and team members Nicola Haines, Karim Hamza, Anja Leonhäuser and Jamie Norden.
7 Since 1998 realities:united has been researching intensively the potentials of fusing architecture and media technologies. Further information is available at http://www.realities-united.de
8 In 2001, realities:united was commissioned by the building's client, the Kunsthaus Graz AG, to develop a "concept for the thorough integration of media technology into the Kunsthaus' architecture." As a result, realities:united developed a broad catalog of ideas, aiming at the creation of an overall "technical character" suitable for the functional as well as aesthetic needs of an institution such as the Kunsthaus Graz. BIX was one "small" part of this overall concept.
9 LED (Light Emitting Diode) technology, etc.
10 Andreas Ruby and Ilka Ruby, "Architecture as a generalist reprogramming of reality" in *ArchPlus*, no. 167, 2003, Berlin: Aachen.
11 spacelab.uk, London, 2000; competition entry for the Kunsthaus Graz.
12 In 2002, the material for the outer transparent and double-curved skin was not yet determined and different materials were being tested. At that time, an argument arose that the outer skin could be constructed from a much less expensive non-transparent material.
13 These software tools, customized for the special needs of artists, were developed in cooperation with programmers from the art scene. This approach eliminated complex translation processes during the production, which would have been necessary if "regular" software companies were commissioned. Software conception: Jan Edler and Tim Edler (realities:united); Tobias Herre and John deKron (thisserver.de). Software programming: John deKron, Jeremy Rotsztain and Peter Castine. Available for download at http://www.bix.at/software.
14 Events are "containers" that can reference local and remote movie files, streaming media files, as well as specific IP addresses authorized to "play" the BIX system remotely during a particular time slot.
15 Besides this administrative function, the BIX Director is also suitable for artists whose work involves mixing multiple source files.
16 Excerpted from an interview with Peter Pakesch in the *coop99* film production, *Kunsthaus Graz, A Friendly Alien*, 2003, Vienna, published by Kunsthaus Graz at Landesmuseum Joanneum GmbH, Graz.

12

THE STRUCTURE OF VAGUENESS

LARS SPUYBROEK

12

THE STRUCTURE OF VAGUENESS

LARS SPUYBROEK

In architecture, flexibility has always been associated with the engagement of the building with events that are unforeseen, with an unpredictable or at least variable usage of space. During modernism that flexibility often resulted in an undetermined architecture, in an averaging of program and equalization, even neutralization, of space. A generalized openness, as we must keep in mind, always has the effect of neutralizing events and being unproductive because the type of space is not engaged in the emergence of events themselves. General, Miesian, openness is only suitable when all desired events are fully programmed in advance, by strictly organized bodies, as in the case of a convention center, a fair or a barracks. It is flexible, open, but it is totally passive. All activity is assigned to the institutional body.

The architecture itself, however, does not engage into the way events and situations emerge; it is indifferent to that. It states that life is merely the effect of decisions that have already been taken behind the scenes, of acts that are repetitions of previous acts, in which intentions are completely transparent.

The ambition then is to find a structure, a tectonics that can absorb life, chance and change, while the structure itself must last and persist over time, to span the unforeseen with the foreseeable. The strategy of the Cartesian grid and the box have always been to average out all possible events, to be general enough for anything. Much of what we do is planned, and much of what we intend is transparent — we script and schedule ourselves — but to engage in the unforeseen does not mean that these are just accidents happening to our agendas. So, the problem of flexibility is not so much "to open up space to more possibilities," but the concept of the *possible* itself. An event is only ever categorized as possible afterwards. The possible as a category lacks any internal structure that can relate the variations; it does not produce variation by itself — it is without *potential.*

The choice has always been between determined functionalism and undetermined multifunctionalism, between early and late modernism, between the filled-in grid and the not fully filled-in grid. But potential is something else: "Potential means indeterminate yet capable of determination ... The *vague* always tends to become determinate, simply because its vagueness does not determine it to be vague ... It is not determinately nothing."[1] Vagueness comes before the situation; neutrality comes afterward. If it comes before, it will neutralize the forces making up the situation. *We must replace the passive flexibility of neutrality with an active flexibility of vagueness.*

In opposition to neutrality, vagueness operates within a differentiated field of vectors, of tendencies, that both allow for clearly defined goals and habits for as yet undetermined actions. It allows for both formal and informal conduct. But more importantly, it also relates them through continuity; it puts them in a tense situation of elasticity. This is, however, not a clean and dry coexistence of two behavioral types as a mere addition or alternation, but more a multiplication, as one comes out of the other and shares the same continuum. To be able to switch between one and the other (in time) we need to materialize their in-between in space, clearly opposing Mies van der Rohe's empty openness and replacing it with solid vagueness.

What becomes evident here is that the architecture of group behavior with all its complex dynamics is directly related to the architecture of the building; a behavior of continuous grouping and regrouping, of solidifying into certain configurations, then suddenly melting and regrouping into other fixed states — a behavioral vagueness paralleled by an architectural vagueness. If the skeletal structure of actions becomes as soft as cartilage and as complex as cancellous bone structure, so does the architecture of the building. We should find ways where the intensive forces dealing with day-to-day decision-making and coping can actually become the formative forces of the architectural structure.

12.1
SoftOffice,
Stratford-upon-Avon, UK
(2000–05),
architect NOX.

SOFTOFFICE

SoftOffice[2] is a building (figure 12.1) where work and play are deeply interwoven. It is a building where both children play and adults work. One half of the building is reserved for very young children to play with interactive environments that are also present on the web. The other half of the building functions as an office where adults work in a so-called "flexi-office" where nobody has his or her own workplace. The office environment is made for both functional, formal conduct as well as for more informal creative conduct, like writing, discussions and presentations.

Intermezzo flexi-offices

In analyzing flexi-offices, it became clear to us that there is a real, daily tension in the effectuation of its usage. This tension between the intended, traditional, static planning philosophies and the viable dynamic structure is actually the force that makes it productive, not the one or the other. In calculating the required surface area of an office for sixty people in very different functions (marketing, administration, online production, offline production, management, origination), one would normally end up with at least 1000 m².

12.2
SoftOffice:
plan.

Yet, within the dynamics of an office culture, if one studies the occupancy rate of spaces that incorporate time-space relationships, one would begin to see a much more differentiated usage over time. Our research gave us three categories of occupancy rates for SoftOffice: 90%, 75% and 35%. The first group would contain people spending almost all of their time behind their own desk. The second group would contain people that also spend a considerable amount of time at other people's desks and at meetings. The third group would contain people traveling around considerably more than the others and spending much of their time in their cars, in restaurants, in hotels or at home.

This dynamic structure allowed us to make the office with 675 m^2. This means 32% of traditional planning was excessive — a miscalculation of the efficiency duly attributed to static thinking. But it is not just the quantifiable side of an office space that has to change; leaving the structure of the office space the same, while reducing it a third in size, would not do any good. Leaving a standard double-loaded corridor type or office landscape (*Bürolandschaft*) does not stimulate the desired extra communication and change of behavior. Practically, then, the spaces and furniture of the office do not need to be designated to a particular person, nor strictly designed for a particular type of work, but in

12.3a–b
SoftOffice:
flexi-office
spaces.

12.4
SoftOffice:
the SCAPE,
an interactive
space for
children.

essence for a state of mind. We set up standard office spaces of general connectivity (flat floor, flat ceiling and general cable access) next to more informal meeting spaces and to the very small capsules for individual work that desires concentration (figures 12.2 and 12.3). The active program is a continuum of both expansion (communicative behavioral types) and contraction (the necessity to shut off, to discuss, meet, write, or shout, etc., either in small groups or alone). The passive program, more a sub-program, is one of bathrooms, cleaning rooms, editing suites and the like.

Intermezzo: children's space

In contrast with the office space, the children's space — the SCAPE — is a space of objects; a field or landscape where a substantial part of the movement is propelled by mock-ups from children's television programs (figure 12.4). Where the adults in the office find a lateral freedom in a fundamentally longitudinally oriented system, the young children's movement in the SCAPE is gravitational and spiraling. They move "around 'n' around" the objects. And they move from one thing to another, from one spiral to another spiral, without any overview: all tension is immediately released and rebuilt again. As most of this is brought about by the iconography of mediated images, the architecture absorbs most of the spiraling in and out as an articulation of the floor surface only. This means the architecture does not need to follow the full movement of the spiral, especially not its rotational nature, just the fact that it is going inward or outward, which has a slight undulating effect on both the floor and roof surface. The rest of the children's movement is produced by a combination of imagery and artificial lighting. Of these there are two categories: objects that are lit and objects that themselves radiate light. The lit objects are less interactive because a pure recognition of the television imagery is sufficient. For other areas a more interactive approach is needed: a zone where the building becomes alive and starts to play with the children.[3]

INTENSIVE DESIGN TECHNIQUES

To map inward going and outward going forces, and to map contractive and expansive forces within one continuum, a networked self-organizing technique is required. An intensive technique means to inform a virtual system, which, during the processing of that information, takes on an actual structure that is a registering of the information. The process has to take on a highly procedural form, like cooking; the instructions are not applied all at once, but one after the other, where timing becomes crucial.

An extensive top-down technique would be satisfied with cutting differently sized holes out of sheet-like surfaces. The closed rooms that are needed for concentration would be subtracted from the surface of communication. In that case it would hardly be possible to create continuity between both states. And continuity is essential if we want tension between states. We are not using the expansive and contractive as finalized properties, we read them as tendencies, as working forces, as formative, not as forms. In the SoftOffice project, we again studied the analog computers developed by Frei Otto.

12.5
Form-finding through analog computing: Gaudí's chain modeling technique.

Analog computing

In the early 1990s, Frei Otto and his team at the Institute for Lightweight Structures in Stuttgart studied what they called "optimized path systems." Previously, similar to the chain modeling technique Gaudí used for the *Sagrada Familia* church (figure 12.5), they had experimented with material systems for calculating form. Each of these material machines was devised so that, through numerous interactions among its elements over a certain timespan, the machine restructures or, as Frei Otto would say, "finds (a) form."[4] Most of them consist of materials that process forces by transformation, which is a special form of *analog computing.*

Since the materials function as "agents," it is essential that they have a certain flexibility, a certain amount of freedom to act. It is also essential, however, that this freedom is limited to a certain degree set by the structure of the machine itself. The material interactions frequently result in a geometry that is based on complex material behavior of elasticity and variability.

In classic analog computing most of the movement is contained in gears, pistons or slots, or often in liquids held by rigid containers but, in the case of Frei Otto's machines, most materials are mixtures of liquids and solids, or they start out as liquid and end up as rigid. In his machines, Frei Otto often used very different materials, such as sand, balloons, paper, soap film (including the famous minimal surfaces for the Munich Olympic Stadium), soap bubbles, glue, varnish, and the ones that we used in the case of the SoftOffice: the wool-thread machines (figure 12.6). This last technique was used to calculate the shape of two-dimensional city patterns, but also of three-dimensional cancellous bone structure or branching column systems. They are all similar vectorized systems that economize on the number of paths, meaning they share a geometry of merging and bifurcating.

In the SoftOffice project, we combined a varnish technique and the wool-water technique (figure 12.6). The varnish technique is a surface-to-line-technique. It is based on the effect that varnish, or lacquer, which is

12.6
SoftOffice: form finding using a combination of the varnish and the wool-water techniques.

12.7a
SoftOffice: the formation of a rubber-lacquer analog computing machine.

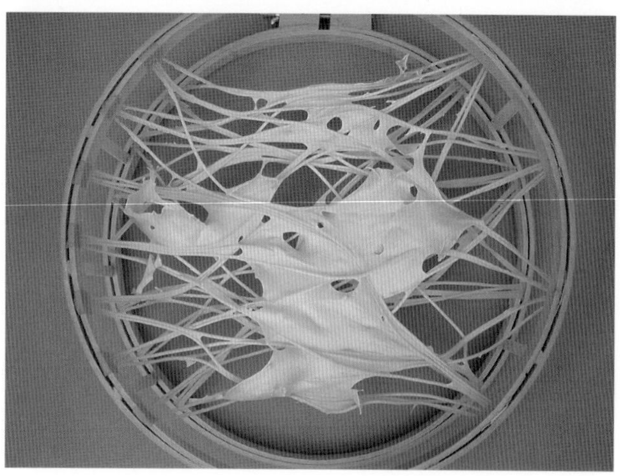

12.7b
SoftOffice: the operation of a rubber-lacquer analog computing machine.

highly viscous, can also later dry up and store information. For instance, one can set up a machine of lacquer surfaces that are being stretched in many directions, break open into holes of various sizes, interconnected by threads of lacquer, which is slowly drying up and hardening over a period of three hours.

The wool-water technique is a line-to-surface technique, where the lines are given beforehand in the form of wool threads, which are set up in a pattern where they are fixed to certain points, then given a certain amount of overlength. When the whole system is dipped underwater and subsequently taken out, threads start to merge (which is a using-up of the overlength), and holes next to surfaces of crossing threads start to form.

Both techniques are fully systemic: all features are formed simultaneously. The holes are not taken out later but they are formed together with the various materializations in the system. The system is calculating everything at the same time, solid and void, during the same process, through thousands of minute iterations, where each positioning is dependent on the formation of another. Order and form are produced, they come about, they emerge during the process. It is a constructivism: a *soft* constructivism, not a Russian mechanistic one. The constructive lines are not rigid H-beams but start as flexible rubber lines that meet up and at the end merge bottom-up into form, into a complex inflexibility. This simply means we use analog computing techniques not just to calculate structural form, but also — on a higher level — organizational form.

The rubber-lacquer machine

In the SoftOffice project, we started with a non-volumetric whole where all the elements were interconnected: a set of lines made up of rubber tubes (of 2 mm diameter) with an 8% overlength, each attached at certain points on a rigid wooden ring (of 450 mm diameter), seven points on the side of the children's space and four points on the side of the office space. From each point there is a rubber tube going to each other point at the other side of the ring, which makes a total of 28 lines (figure 12.7a). We doubled this system: two wooden rings each with 28 tubes that not only connect one side to another, but also one ring to another. This system was then dipped into a very liquid lacquer, analogous to the wool-water technique. But while the wool-water model is always flat, the two wooden rings can be separated during the hardening process. Instead of having the holes and mergings in a flat configuration, we can now calculate curvature of the rubber tubes together with the intermediary curvature of the drying lacquer in a spatial configuration. The separation of the wooden rings during a three-hour procedure is analogous to the splitting of the floor and ceiling (figure 12.7b).

So, while calculating programmatic forces — mental states — we are also calculating structural forces. There is complete vagueness: never fully column, never fully wall, never fully floor (figures 12.3 and 12.4). The system is negotiating everything with everything without resorting to equalization. Now, what becomes most prominent in this system is that there is both an expression of rigidity and of flexibility. We found a methodology that allows us to calculate the in-between, to be able to vary between the two states, regardless of how opposite they are. While all flexibility is expressed in the middle of the system, the rigidity is produced close to the wooden rings, at the edge of the system. Type is at the edges, diagram in the middle; full bottom-up in the middle, full top-down at the edges — spatially: a spongy porous structure in the middle zone, clean separation of floor and ceiling at the edges (with columns in-between). In the SCAPE, this clean separation means the hall-like tendency of the structure; at the office side, four separate 'fingers' with gardens. It is now the in-between that becomes operative: it is not just a Cartesian choice, it is an actual sense of tension, a material state of in-between that is internalized, that becomes effectuated in daily behavior and functioning.

Soft rigidity

The first step in the previously described process contains only geometry, no materiality; the materiality then takes over during a stage of reshifting and the procedure comes to a halt in a state of full geometry again, but a geometry that is now not imposed on a material but is the result of material interactions. It starts out explicitly Euclidean, but it does not finish as such because at the end there is no clear segmentation of dimensions any more. While we could call the first step of the system a geometrical surface, a system where all directions are equally present, the final stage of the model is much more complex because it consists of patches of crossings, mergings and holes. The crossing patches consist of two dimensions, which means that, in these areas, many directions are still available in the system — many lines keep on crisscrossing each other, similar to the initial state. The merging patches consist only of one dimension, where the system takes on one single direction — many lines stick together to form a main artery. And the holes, of course, are areas where we lose all dimensions and no directions are available anymore.

While the first stage consists of homogeneous tiling, the last stage consists of heterogeneously-nested patching. The end result (figure 12.7b) is based on looseness, but is itself not loose and not weak, but rigid and completely tight (when attached on an open ring it comes out of the water straight and horizontal). It is a strategy of flexible, individually weak elements cooperating to form strong collective configurations.

What emerges is a complex or *soft rigidity*, which is very different from the top-down, simple and *frozen rigidity* of the first stage. We should therefore resist the idea that the first stage is a rigid order and the end result is just a romantic labyrinth or park. Actually, the arabesque order of the end result is as rigid as the first stage of the grid, but much more intelligent because it optimizes between individual necessities and collective economy. Yet it is not an easily readable and clear form of order, but a *vague order;* it is hardly possible to distinguish between surface areas, linear elements and holes. Surfaces can function as linearities, lines can cooperate in surfaces, and holes can exist at all scales. Everything between the dimensions is materialized. Although the dimensions are clearly singularities arranging the system (the mergings into thick lines are like the ridges of dunes, which orient the sand surface to the wind forces), it is continuity that makes them emerge — although the order is vague, it should nonetheless be considered very precise because nothing is left out. There is no randomness; there is only variation.

The interesting feature of this system is that it is in fact structured by holes; the nesting of holes is the driving force behind its formation, while architects are always trained to think that holes are, in the end, subtracted from a system. This machine does not operate on subtraction or addition, but on multiplication, in the classic sense of early systems theory, which states that a whole is always larger than the sum of its parts. Here porosity is an emergent property. The first stage (figure 12.7a) is basically *drawn*, contrary to the end stage (figure 12.7b), which is processed by a machine, i.e. *calculated*. All effects that coexist in the final result, all the curves, all the mergings, all the holes are interrelated; nothing can be changed without affecting the arrangement of the whole. All lines are mobilized simultaneously, in parallel, while drawing is serial; one line is drawn after the other. A drawing is always created in the visual field, while the analog machine follows a partly blind and informational logic where the image is the end product of the process. And although this technique should be considered as a hybrid of the top-down and the bottom-up, the drawn and the generated, its intelligence lies in the fact that nothing is "translated;" the drawn is not "translated" into the real. In itself it works in one-to-one scale — in that sense it is not even a model. This *direct proportion* is one of the main features of analog computing, which does not simulate by numbers but by an empirical rescaling of the real.

The organizational and informational stage is material, not immaterial, as is so often put forth. It is the material *potential,* the material distributed intelligence which sets the machine in motion — a transfer of water-turbulence to wool-curvature. Then it is the stickiness, the hairiness and the curvability of the wool thread together with the cohesive forces on the water surface that bring it to a halt again and inform the end result. It is an intensive technique within an extensive system, and though the quantities (surface area, etc.) are given beforehand, the quality emerges through the interaction and multiplication of different parameters. Generally, the intensive is a deformational property (like heating), but here it also becomes a transformational property (like boiling): the threads restructure and reorganize to "find form." The system as a whole passes a critical threshold. The degrees of freedom of deformation, which are more like extensive movements within an internal structure, become intensive, qualitative changes of "that" structure.

Wet grid versus dry grid

The classic regular, Greek grid is a system that separates infrastructural movement from material structure. Simply put, the structure is of a solid, while the movement is of a liquid. We must consider the orthogonal grid as a frozen condition because its geometrical state of homogeneity relates directly to a *material,* crystallized state of frozenness. Frozen states are simple states and, of course, these have been the first to be mastered by the *geometers,* but to understand complex states we need to develop complex geometries. Generally, we are taught to think that geometry is the higher, the more abstract and pure form of materiality, which is a misconception because although geometry urges for the necessary exactitude, it is totally imprecise. Any geometer comes after the event, when everything has dried up and, therefore, he can only be dealing with the extensive state of the material, taking up length, width and height.

The wet grid, Frei Otto's grid, is one in which movement is structurally absorbed by the system; it is a combination of intensive and extensive movement, of flexibility and motion. The geometry does not follow the event, geometry co-evolves with materiality; it is generated through analog, wet computing. One could call the organization of the final stage wet and its structure dry. While it itself is not moving anymore, it has attained an architecture of movement. In this sense, movement must be viewed as information, as pure difference, because we all know that when "information" does not cause any change it is superfluous. It simply did not "in-form," it did not enter the form. This means movement in itself is not enough to be called information, it must be internally processed as a (temporary or permanent) transformation. The physical displacement of movement must be processed as a structural change.

Basically, emphasis on movement as deformation is merely indexical and is meaningless when not resulting in structural transformation. The freezing of movement becomes merely *traces,* momentary stoppages of a bygone present; they are not, however, structured through time, they are not *paths* which allow for movement to be rerun over and over again, and slowly condense and evolve. On the other hand, they are not *roads* either, which with their exact distinction between surface and line prevent the system from reconfiguration and adaptation. Each state of path-forming should function as the analog computer of the next one. There should be enough solidifying for registering and there should be enough plasticity to enable changes. This brings the optimized path systems of Frei Otto close to contemporary multi-agent computing devices based on ant colonies with their pheromone distribution.

For a real-time, analog computing model we need two things: first a system that is internally structured (or else it cannot process information) and, second, external flows of information. This simply means there are always double states, simple states and complex states, coexisting in gradation. Higher states of information can only happen in lower states of information, they coexist hierarchically but within a continuum. They do not exist next to each other — the generic and the specific share the same continuous, topological space as do the standard and the non-standard. One is always engulfing the other: we need to start from a state of equilibrium that already contains information through its structure, create disequilibrium to increase the amount of information, and then we need equilibrium again to memorize it.

The brilliance of the Frei Otto's model is that the flexibility is taken literally and materially, that the real movement of water flow becomes the abstract movement of wool-structure, which results in a coherent language of "bending," "splitting," "curving," "nesting," "aligning," "merging" and the like. All arabesque figures in the final state of the model immediately relate to complex configurations.

obliqueWTC[5]

Undeniably, the skyscraper is the most successful building type of the twentieth century. Its generic reductionism, however, its passive stacking of human behavior, its manic monoprogramming will and should become obsolete and, as a type, it will have to be rethought, making a new evolutionary step of the megabuilding possible. We should try to find urban strategies to deal with the "Huge," with global forces working on local situations. We should find ways to work against the homogeneous and find other ways that are more open to life, the changes of life, and the unpredictability of life.

In rethinking the mega, the Huge (which is slightly related to the "High"), we should be more concerned with the structure of the Huge than its size. Developing techniques to heterogenize the Huge, without just simply chopping it up or collaging it together, means we have to reconsider the design techniques. Moving away from top-down towards bottom-up techniques — especially when dealing with superlarge structures — becomes absolutely necessary.

In developing our design for the rebuilding of the World Trade Center (WTC) in New York, we reused the old wool thread modeling technique invented by Frei Otto and described in previous sections. We used one wool thread for each core of the destroyed or damaged buildings on the former WTC site. In an inverted model, the wool threads hang straight down under the sole

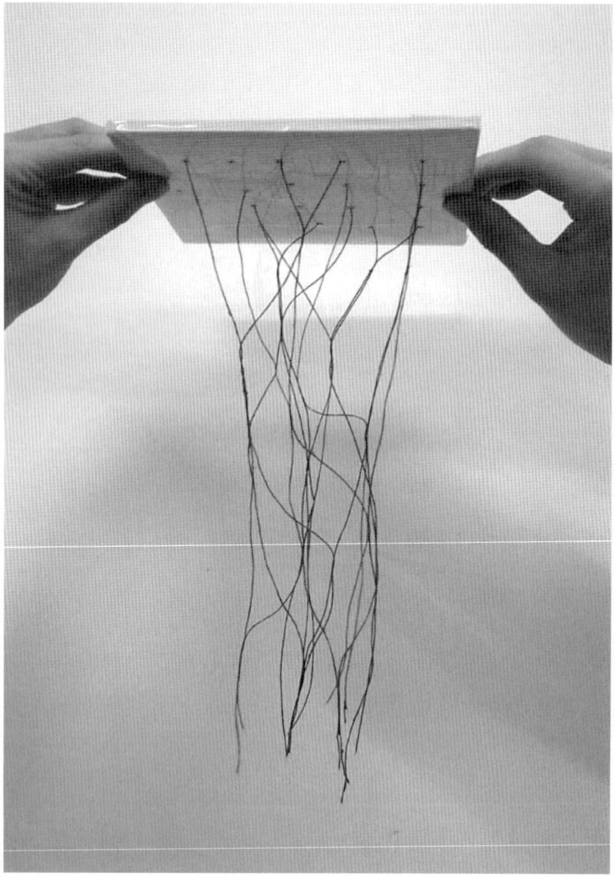

12.8
Form-finding: World Trade Center, New York (for Max Protetch Gallery, 2001), architect NOX.

influence of gravity forces. When dipped into water and taken out again, all threads reorganize themselves into a complex network[6] (figure 12.8), comparable to bone structure. The structure is no longer formed by a simple extrusion of a plan but instead self-organizes into a networked megastructure, where the whole is larger than the sum of its parts.

We thicken each of the wool threads into a lean tower that merges and splits up as it moves upwards (figure 12.9). This enables the structure to comply with the New York zoning law that only allows high buildings to occupy 25% of the total site surface area. In this case, however, the 25% is always positioned somewhere else, making it both into one single megabuilding with many (structural) holes in it or many thin towers that cooperate into one large structure. The towers sometimes act as a bridge, sometimes as a counter-structure for another one, and sometimes free themselves to become a smaller sub-tower. Most of the loads are carried through the honeycomb steel structure of the surface, helped by an interior column grid, which follows the diagonals of the towers similar to Louis Kahn's design for the Philadelphia City Hall from 1957. Also, the elevators form a highly complex structure of diagonals where, at some platforms, more than five or six different cores come together to form larger public areas. It is this network of elevators which makes the building not just a new type of tower but more like a new type of urbanism. The elevators become an urban extension to the subway system: a punctuation of the street by a technological system to intensify its public functioning. Generally, all interactions of a Manhattan block (with its programmatic diversity that should at least be rivaled by this new building) only happen on the street, while all buildings blindly tower away from that level into a non-interactive side-by-sidedness. Here we re-network the street into the tower. We read the wool-thread diagram both structurally and programmatically, where the structural 'diagonals' become a reemerging of Virilio's *oblique:*[7] lateral, horizontal street forces are multiplied with the vertical stacking model of the skyscraper, resulting in an oblique tower.[8]

12.9
World Trade Center, New York (for Max Protetch Gallery, 2001), architect NOX.

A SOFT CONSTRUCTIVISM

The techniques invented and suggested by Frei Otto have been very diverse, varying between the application of already invented techniques to ongoing projects and more fundamental research into material form-finding. Not surprisingly, his optimized path system machine is quite unique within the whole of his research because he rarely bothered with horizontal structures. Essentially, his research was in the complexity of the elevation, the structure and not the plan.[9]

Patterning effects and configurational, emergent effects happen on all stages, both in the plan and in the elevation. Instead of following the plan–floor/extrusion–wall method, we should opt for a method where elevation and plan become more intertwined and coevolve into structure. For centuries, the order within the design process has been: first the plan (action), then structure at the corners (construction), which at the end is filled in with walls (perception), where the latter two have been part of the splendid Semperian distinction between tectonics and textile.[10] Our agenda should be to short-circuit action, perception and construction. Having weak textile threads team up into rigid collective configurations is a direct upgrade or inversion of the Semperian paradigm. But they should be three-dimensional from the start; plan threads can twist and become wall threads.

All these techniques already exist in textile art, where complex interlacings occur in crochet, weaving and knitting. The art of the arabesque is as old as architecture; it has just never been conceived at the scale of the building. And that is certainly because of technological reasons — the arabesque has always been accommodated by manual labor while the straight extrusion was necessarily associated with standardization and industrialism. Clearly that is changing with non-standard architecture.

We should be careful though not to mistake the non-standard for "free-form architecture," for the amorphous or even the streamlined; we should strive for a rigorous non-standardization, rethinking repetition within sets of variability, rethinking structures within ranges of flexibility. The more we move towards the non-standard, the more articulation must become an issue. If there is no technology of design, a technology of manufacturing becomes nonsensical. With machines under numerical control, we need the design process itself as an informational procedure also, with clearly stated rules and scripts to generate a structure of vagueness.

We have argued here and before that starting with the soft and ending with the rigid will offer us much more complexity in architecture. And here we are not referring to Venturi's linguistic complexity (of ambiguity)[11] but to a material complexity (of vagueness). Obviously, the science of complexity has produced many diagrams of the soft, and these have often been dropped onto rigid architectural structures or typologies. Although *deconstructivism* proved to be successful in breaking down most of the top-down ordering tools we were used to in architecture (contour-tracing, proportion, axiality, etc.), it proved to be incapable of instrumentalizing complexity itself as a tool that was material and architectural. It understood every act of building as an implicit counter-act, as a negation — meanwhile, the engineers silently repaired it.

We should understand all objects as being part of a process of emergence; *the made as being part of the making, not the unmade.* Our goal must be *constructivism,* or emergence, and anything that emerges should coemerge; the way we see is emergent, the way we move around, the way we act in relation to others, to our habits, to our memories, all these emergent patterns should coemerge with its material structure.

This makes our agenda one of a post-industrial constructivism, a non-standard constructivism. All behavior is material, all structure is material. "How do we orient, how do we feel, how do we group or ungroup?" all these questions should be posed simultaneously, together with "how does it stand up?" There have been many attempts to borrow "images of complexity" that were fed into either circulational, formal or structural diagrams; Klein bottles, weather maps and the like are interesting but are not enough. We should create complexity by feeding them into each other. We should feed circulation into structure, feed structure into perception, and feed perception into circulation. It does not matter where we start as long as we are looping a flexibility of action (affordances) into a flexibility of structure (vagueness) into a flexibility of perception (atmosphere), looping non-standard behavior into non-standard structure into non-standard architecture.

NOTES

1 C. S. Peirce (ed. Peirce Edition Project *et al.*), *The Essential Peirce: Selected Philosophical Writings*, vol. 2, Bloomington, IN: University of Indiana Press, 1998, pp. 323–324.

2 Shop, interactive playground and headquarters office for Ragdoll TV Productions in Stratford-upon-Avon, UK (2000–05), architects NOX (Lars Spuybroek with Chris Seung-Woo Yoo, Kris Mun, Florent Rougemont and Ludovica Tramontin).

3 The area is what we have called "Glob." Glob is a world designed by globally networked children. Glob is both present in the building and on a website. Glob is a "living organism" (some of its responses are calculated with genetic algorithms) that has the special ability to interact with children. Glob will grow, love and, of course, play. Glob creates an experiential environment for the children that touches upon their senses and humor. Together they will create drawings, music, stories and love.

4 Frei Otto refers to this on many occasions as *Form Findung* — finding form or taking form. Together with Bodo Rasch, he published a book with the same title (*Finding Form*, Munich: Edition Axel Menges, 1995). The terminology in German was not used before Frei Otto coined it. One should frame the *formgebung-formfindung* debate in a German discussion going back far earlier, especially the one during the late eighteenth century on Preformation and Epigenesis. For a more elaborate discussion see Detlef Mertins, "Bioconstructivisms" in Lars Spuybroek, *Machining Architecture,* London: Thames and Hudson, 2004 (forthcoming).

5 NOX (Lars Spuybroek with Chris Seung-woo Yoo and Kris Mun) for the Max Protetch Gallery, New York, 2001.

6 The cohesive lateral forces of the water are added to the gravitational system.

7 Paul Virilio and Claude Parent, *Architecture Principe*, Paris: Les éditions de l'imprimeur, 1997.

8 We included a Memorial Hall inside the building. High up in the structure several floors are taken out to form a large open space that gives an open view over the city, but the space will also be visible from most areas around New York. The hall should not be a monumental petrification of mourning but should be a projection space where visitors can interactively request for home videos, photographs and websites of the all people lost, and meet them.

9 Frei Otto was always invited to cooperate with architects that had already developed the plan, and his contribution was subsequently in the typical engineering stages, afterward.

10 In *Die Vier Elemente der Baukunst* (*The Four Elements of Architecture*, London, 1851), Gottfried Semper made his famous categorization of the elements of architecture through a material classification: a. earthwork/foundation; b. tectonics/wooden poles at the corners; c. textile/wall as an infill of the wooden poles; and d. the hearth. In the same vein I always add: floor/action, corner/construction, wall/perception, and fire/sensation. Semper's categorization is based on his seeing of the Caribbean Hut during the Great Exhibition in London where the structure consisted of load-bearing columns of wood on the corners, with walls made of the non-load-bearing element of woven textile. Although we here fully underline the implicit notion of surface and textile as the main constituent element of architecture, we can basically consider textile now as a tectonic element in its own right, especially while referencing to Frei Otto, not so much to his application of textile in tent architecture but through his methodological use of textile in the analog computing of form.

11 Robert Venturi, *Complexity and Contradiction in Architecture,* New York: Museum of Modern Art, 1977.

13

PERFORMATIVITY: BEYOND EFFICIENCY AND OPTIMIZATION IN ARCHITECTURE

ALI RAHIM

13

PERFORMATIVITY: BEYOND EFFICIENCY AND OPTIMIZATION IN ARCHITECTURE

ALI RAHIM

Technical articles of research tend to use the term "performance" but rarely define its meaning. How does the performance of a computer relate to the performance of a material system? There is no general definition of performance.[1] The precision of each description comes from its local context. The performance of a technology, for instance, refers to its technical effectiveness in a specific evaluation or in a set of applications.

Technical disciplines are constituted around devices conceived as essentially functional and therefore inevitably oriented towards efficiency. This obsession with efficiency is prevalent throughout society and is reflected in the design of many devices and systems, but it has not demonstrated an understanding of technology as it has existed historically, as it exists today, and as it may exist in the future. Technology is not merely technical; it is an active and transformative entity resulting in new and different cultural effects.[2] Technology in this sense is not an efficiency-oriented practice, measured by quantities, but a qualitative set of relationships that interact with cultural stimuli, resulting in behaviors, some of which are techniques.

Our work uses techniques of design that are process-driven and that produce these transformative effects in cultural, organizational, technological, social and political production. Techniques have always contributed to the production of human and cultural artifacts, but their refinement and acceleration after the industrial revolution has emerged as the single most important element in the evolution of cultural endeavors.[3] Techniques influence the design of objects, which, in turn, produce effects that influence human behaviors and produce new techniques that affect an object's technical performance. This results in the selection of new possible objects which generate further cycles of development.

PERFORMATIVITY

The path of evolution produced by a cultural entity — an object, a building, a company or a career immersed in its context — produces a distinct lineage[4] as a result of its propagation. Each lineage exists indefinitely through time, and may be selected in terms of its performance. Although different performance paradigms are separate, they inevitably emanate and influence each other. This selection always constitutes an emergent and decentralized process in which many actors are interrelated through complex feedback loops. This process requires cognition, is innovative, and leads to a restless proliferation of new effects. This transmission across lineages is, as Stephen J. Gould writes, "the major source of cultural change."[5]

Animation techniques enable us to inhabit these potentials in time and cross these lineages. Our design process reacts to external stimuli and transforms a situation through feedback between a subject and the environment and between architecture and its milieu. The material, organizational and cultural change that occurs as a result of this perpetual feedback and two-way transfer of information is *performativity*. Here, models developed in one research paradigm can be appropriated by another. These paradigmatic readings not only have the ability to generate, describe and evaluate performances, they also cite and recite them, breaking apart the evaluative forces that bind together their discourses and practices. More importantly, they can recombine and reinscribe these forms, deploying them elsewhere while incorporating, ignoring or reevaluating their values in other ways.

Performativity always has the potential to produce an effect at any moment in time. The mechanisms of performativity are nomadic and flexible instead of sedentary and rigid. Its spaces are networked and digital rather than enclosed, and its temporalities are polyrhythmic and non-linear. Performativity produces new subjects of knowledge, hyphenated identities, transgendered bodies and digital avatars. The performative subject is fragmented rather than unified, decentralized rather than centered, virtual as well as actual.

13.1
Plan view showing different inflections across the inhabitable surface of the Leisure Generator (2003), architect Ali Rahim/CAP.

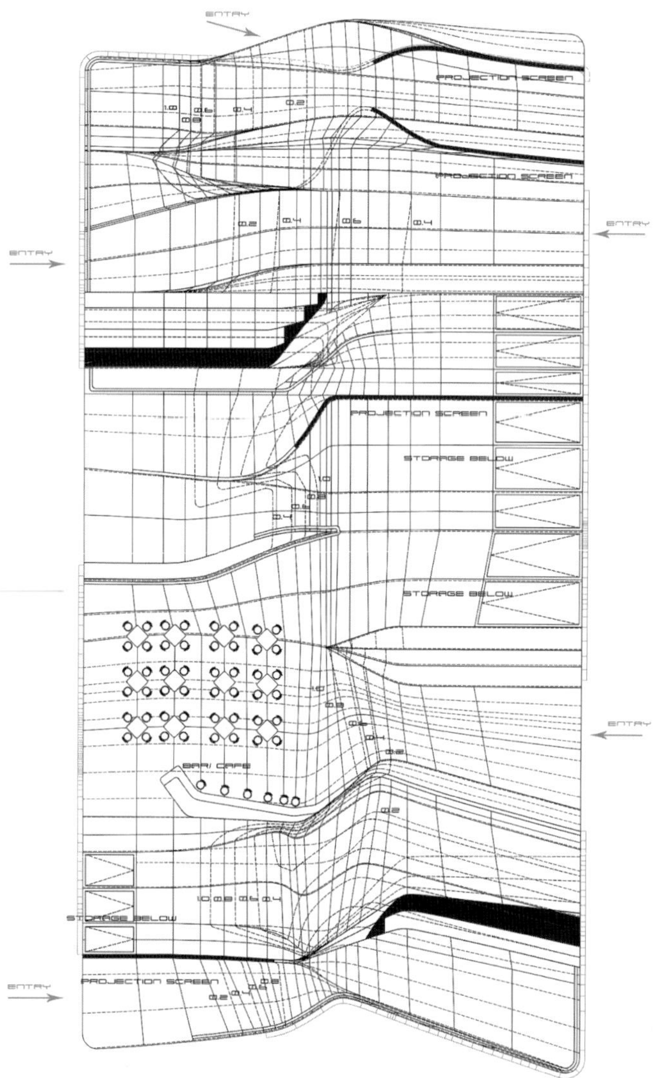

In our winning competition entry for a Leisure Generator, sponsored by the Greek Government in Association with the Hellenic Ministry of Culture for the 2004 Olympic Games, our design strategy is to sample, providing the greatest range of potential effects that different cultures might produce in interacting with a surface and one another. The design celebrates the heterogeneity of all world cultures by intensifying different leisure activities and cross-pollinating sporting, social, recreational and domestic leisure categories during different periods of the day (figure 13.1).

We allow for many leisure spaces to occur simultaneously in order to stimulate unforeseen relationships that might emerge through the feedback between bodies and the inhabitable surface of the project. Here, different cultures are homogenized by the body rather than by color, gender or religion. Some bodies may sit on a surface while others may use the same surface for laying or reading. Individual users may influence the person next to them, affecting the way that groups form and interact with the surface. These relationships become more direct at the more particular geometries; for example, stairs can be used for walking, sitting or sunbathing (figure 13.2).

The structure is redundant and built in an accumulative manner. It eliminates thresholds and moves them towards a gradient where one can have different levels of opacity in one surface or different lighting qualities within a single space. Opaque, translucent and transparent effects can occur in one surface in continuous variation. The conflation of structure and material combine to change relationships and recombine to produce effects that are guided by the formation. Here, the detail is everywhere. This non-reducible geometry allows the subject to move through this field with a degree of ambiguity, avoiding conventional norms of expression and prescribed interpretation.

Performativity influences the outcome of habitational, material and ambient effects perceived by users and the effects they have on their milieu by reconstituting our sensibility of architecture (figure 13.3). These formations have no bounded limits in figure or ground, building or landscape, inside or outside, public or private, but provide a continuous gradient between extremes and a rich palette of possible interactions. This constitutes an unfolding field of

13.2
Leisure Generator: detailed plan view showing the accumulation of the structure through time.

social discourses facilitated by geometric articulation. Our intention is to catalyze interactions between the formation and the subject by opening the work up to a wide field of possible orders rather than a single definitive one.

The emergence of the computer is another example of this potential. It is clear that each technological lineage embodied in the computer encompasses the contributions of scholars, philosophers, visionaries, inventors, engineers, mathematicians, physicists, and technicians. Each lineage is stimulated by vision, need, experience, competence, and competition. As these lineages develop through time, they organize effects already in existence, such as the use of machines and automation. They organize advances made in symbolic logic and in science and mathematics, which become feasible with the invention of numerical patterning systems. These factors are impacted by different intensities of economic, commercial, scientific, political and military pressures, crossing the technical threshold and emerging into the temporally organized technological formation that is a computer. The computer here is an effect of hybridization across these lineages.

If we were to view these non-linear organizational processes as fixed in space and time, the resulting objects would be severely limited and would strain to represent meaning through formal expression. This object type is passive and defined only by its material attributes which are linear and causal. Such an object is static and only has the capacity to produce predetermined effects. To avoid this stasis, we must view the object in its context and understand it as a part of a continuous temporal process. For example, several types of machines, including

13.3
Plan of landscape and building at Variations, London (2002), architect Ali Rahim/CAP. The form of the building and landscape are developed simultaneously as one is an intensification of the other.

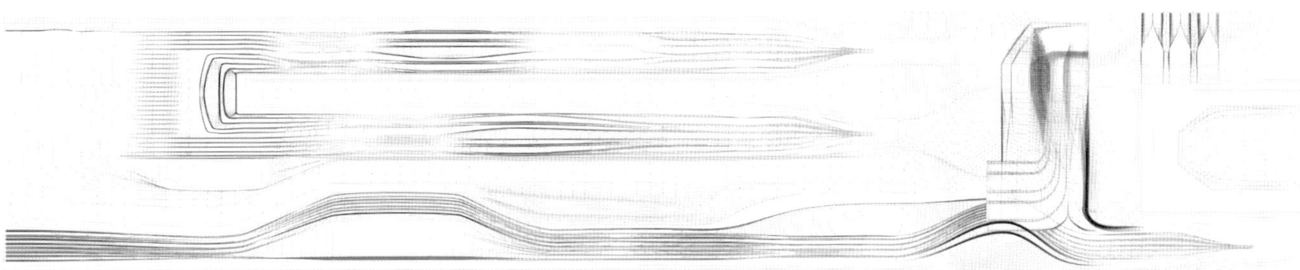

13.4
Variations: entry — there is no threshold between building and landscape, but a continuous gradient in-between.

13.5
Variations: an exterior view of the building showing its formal variation.

tabulators, sorters, transcribers, and verifiers, were cross-affiliated to produce the computer.[6] If each machine is viewed as being fixed in space and time, then the sorter, for example, would become a better sorting machine and not part of the organizational process that influenced the emergence of the computer. This cross-affiliated process operates from within, is influenced by its environment and has a potential to spontaneously self-assemble and produce effects which are qualitatively larger than initially anticipated.

The computer is instrumental in affecting its users. They are influenced by their interactivity with it and, in turn, impact their environment. This feeds into a cycle of productive emergent effects which influence our cultural milieu. This process is temporal and each point in time becomes a potential generative juncture through which temporality is established.

A desire to study the behavior of matter in its full complexity brings with it an ordering of the world where material systems are understood to be similarly cross-affiliated and in a constant state of flux. The analytical approach to design, which studies a site or project only from the top down, misses the properties that emerge from complex interactions between parts, as in the modernist notion of the diagram as building *parti*.

Our work seeks to harness the potentials of performativity by using temporal techniques to simulate the indeterminable unforeseeable influences and inevitable unforeseen events and accidents that invigorate dynamic environments. We study the behaviors of animations over time, analyzing both quantitative changes and qualitative rates of change. We guide and shape formations from within this process, using the computer's ability to generate and regenerate possibilities iteratively, adjusting and fine-tuning relationships between matter and use.

The resulting formation is itself new, with new properties that do not belong to the old. Just as water is not a property of hydrogen and oxygen, and a tornado is not a version of the winds that precede it,[7] inorganic matter is much more variable and creative than we ever imagined. One way to fully participate within complex environments is to develop a spatial-temporal formation that negotiates the real-time response of the system that helps generate it. Here the traditional axis of hypothesis, analysis and intervention gives way to perpetual feedback between analysis, intervention and exchange with the environment time and time again (figure 13.4).

13.6
Variations: the inflections in the surfaces influence to different extents how the individual uses each space, and how this affects group use.

This dynamic exchange between intervention and environment is studied in Variations, a weekend residence for a fashion designer in London (2002). The project is also a venue for fashion shows held seasonally for her best clients. Here, the notion of a quantifiably efficient design looses its meaning (figure 13.5). There is no optimal fixed solution once and for all, as these criteria change over time. Habitational potential emerges from the spatial and material attributes that constitute the project's form. No space is described as a function of its use or by event, but of the qualities it contains.

Animation techniques were used to simulate performative effects (dining, sleep, spectatorship, work, etc.) at different scales through an iterative process (figure 13.6). One scale examined the feedback between building and environment

13.7
Variations: three speeds of circulation are brought together through an inflected surface of different extents.

(drainage, vegetation), the second between the events schedule and the form of the building and landscape, and a third scale studied shifting scenarios that emerge on a daily, weekly and seasonal basis. The actual habitation of the space emerges through negotiations with the surface by users and may vary with different social situations (figure 13.7).

Several tendencies emerge through this interaction. These are both specific and charged in certain places and primarily unaltered in others. Patterns of use include crowding or dispersion and always shift between extremes in the same space. Redundancy within the structural strategy of the project provides for greater formal variation, as does redundancy in the surface articulation. For example, in the openings, we studied the patterning of light through different times of the day and year (figure 13.8). We wanted to intensify the gradient between different conditions, such as light and dark, smooth and rough, loud and quiet. We also provided for multiple routes through the project, some fast, others slow, with the intention of giving the form greater potential to be used differently.

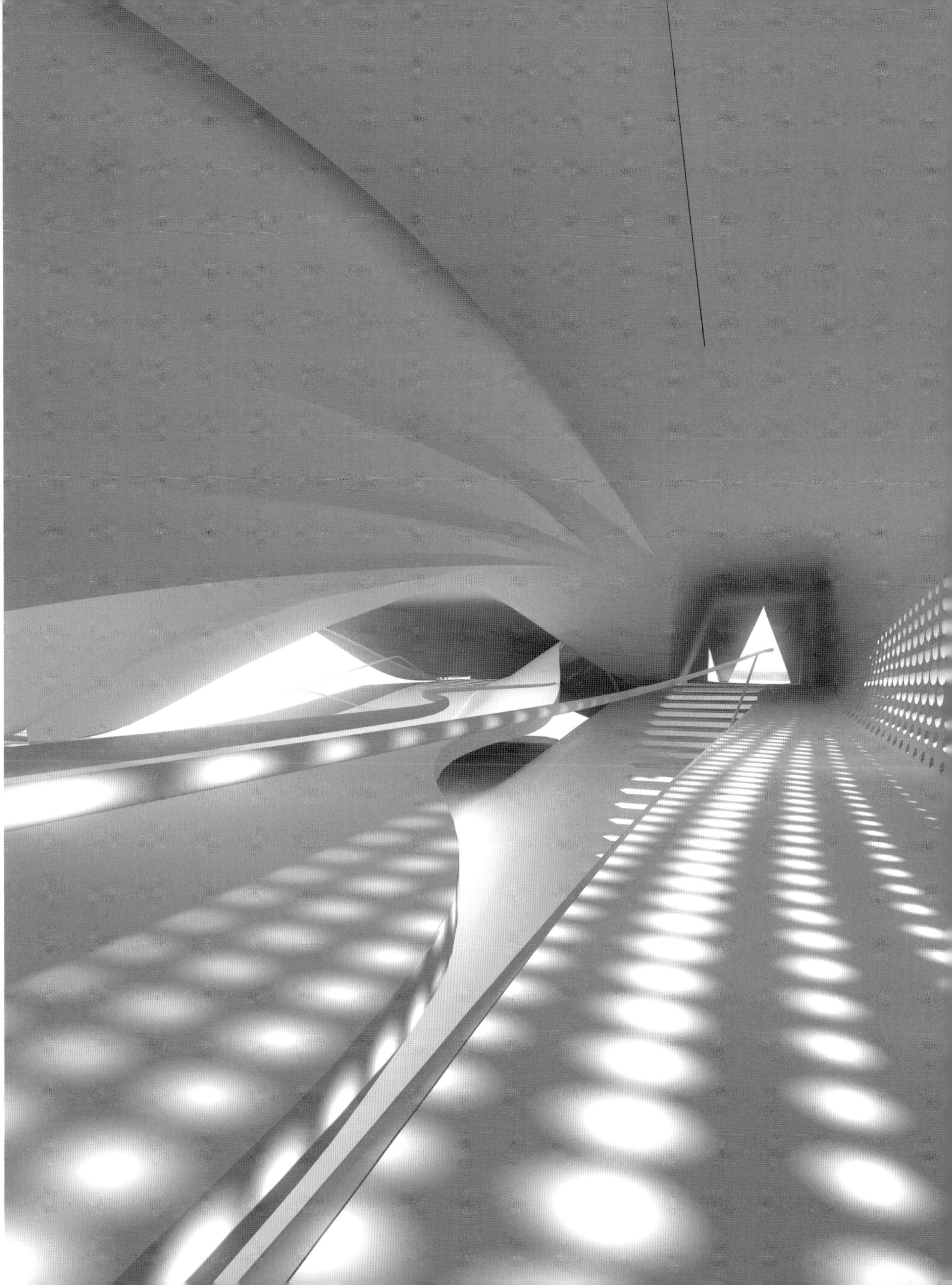

13.8
Variations:
the lighting
gradient.

These potentials are not achieved through efficiency. Instead, they are products of performativity where design, manufacturing and optimization are seen not as static elements within the development of a building but as participants in a dynamic model and as part of a temporal process. This shifts optimization and manufacturing from being efficiency-based practices, concentrated on quantities and the production of static objects, towards being conditioned through their development and evolving possibilities for new spatial-temporal qualities and experiences. These effects are seamless between environment, user, formation, structure and material distribution.

BEYOND EFFICIENCY AND OPTIMIZATION

Optimization often points towards the most efficient or structurally minimal solutions. While non-linear analytical software is available, it is typically used on a project post-design to make sure a structural and mechanical system is viable. This model of optimization always seeks to maximize the efficiency of a form, searching for the purest solution in an attempt to do more with less material, move air better, etc. How much more efficient can a design get? How much thinner can a cable get? We can already imagine that due to advances in material science, in fifty years the smallest tensile member might be as thin as a hair, but the reduction of material quantity to its minimum, or the reduction of a form to its most efficient structural expression, would fail to realize the relational and formal potential of the new materials. By all evidence, these techniques of efficiency and reduction from the top down have reached their limit and, in all likelihood, they will be unable to produce new possibilities for inhabitation or to generate new effects.

The software being developed in advanced engineering offices is able to generate and test forms dynamically. This software is pointed in the correct direction but its focus needs to shift from structural and mechanical optimization towards a way of evolving new possibilities for spatial-temporal experiences and material potential in design and manufacturing. The manufacturing industry needs to become more flexible and adaptive, and shift towards a model of performativity. This would entail evolving a building through performativity and simultaneously testing its effects in different social settings. This would allow designers to develop and test their formations immediately in the pursuit of producing effects and environments that have habitational potential. It would also forecast unpredictable conditions and would inform us in real-time about our formal, material and ambient effects.

Incorporating manufacturing within this model would shift production processes from mass production and mass customization to one where the dynamic evolution and production of non-prescribed possibilities is possible. Some manufacturers, like Nike in its NikeID line, are shifting towards personalization as a manufacturing process, but they still rely on postmodern collage techniques by controlling the number of possible outcomes. They simply provide a wider matrix through variety that customers can select from instead of a more fluid variability.

We can look towards the automotive industry for inspiration and a possible direction for developing techniques based on greater temporal redundancy. Here, there are many scales of redundancy. For example, crash testing of the automobile chassis is done both virtually and physically, and in as many scenarios as possible. Automobile designers employ an iterative process and come across effects that they can predict and others they cannot foresee. These effects are generated by the variables that produce them and it is this outcome of effect that highlights these virtual techniques and reveals the potential for them to be applied in architecture and other design disciplines. This potential arises from the fact that the total environment is incorporated within the software in order to produce a dynamic testing ground for material and effect. Virtual crash testing is very specific. Speed, material properties and environmental conditions are all coded within the software, which uses magnitude of force and direction to anticipate deformations in the vehicle and injuries to the passengers. This allows for highly precise calibrations within an evolving performative feedback structure that has the capacity to generate and test emergent effects on users throughout the design and manufacturing processes.

13.9 Catalytic Furnishings: lifestyle diagram.

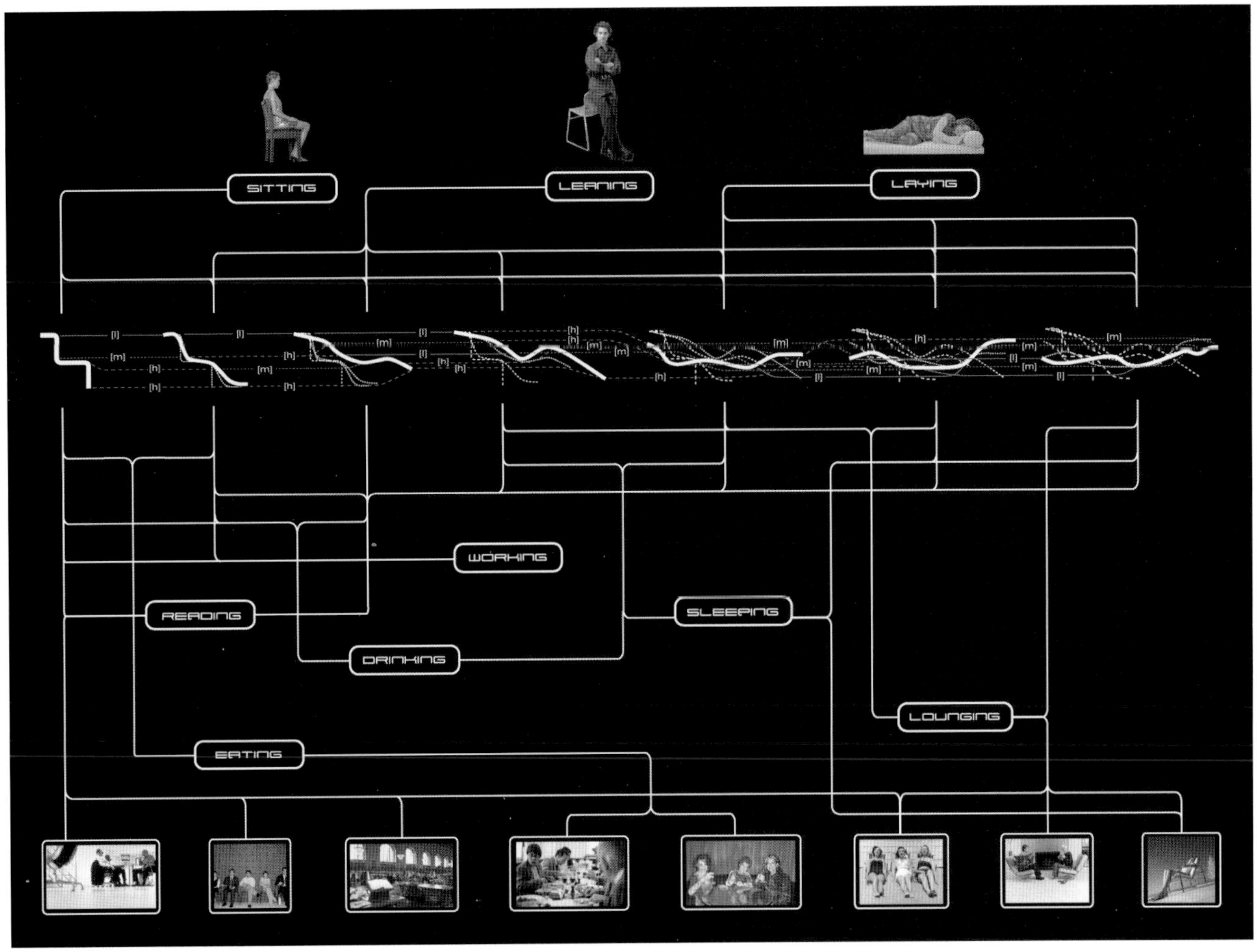

Our Catalytic Furnishings project (2003) points towards the potential of a flexible manufacturing technique that evolves through the design process. We were interested in the way the body acts in ways that we anticipate as well as ways it might act differently depending on interactions with the surface of a furnishings element. We aimed to charge a simple material surface with varied potential for occupation and use.

The project mixes and samples several different lifestyle types and uses them to understand primary body performances on different orientations and surface qualities (figure 13.9), from rigid vertical surfaces for leaning to softer more horizontal surfaces for laying, and working to compressible horizontal surfaces for sunbathing. The heterogeneous geometry of the pieces tends to destabilize simple division between individually occupiable spaces and necessitates an ongoing territorial negotiation between simultaneous users. Users input variables of their lifestyle into software that deforms and shifts a structural cage (figure 13.10) and the manufacturer has the ability to dynamically vary the form of a mold in real time (figure 13.11). This variable mold system we are developing, called FlexiMold©™, allows for cross-affiliations of different body performances to be registered. Through a performative process of selection, users are able to create their own customized object that responds to their particular lifestyle.

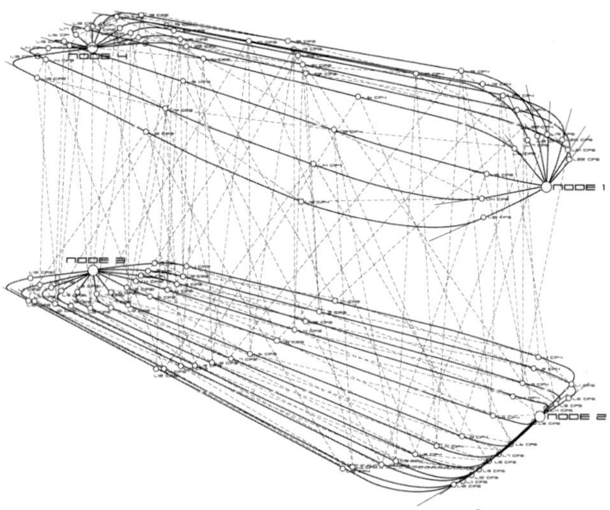

13.10
Catalytic furnishings: cage deformations — users input variables of their lifestyle into software that deforms and shifts a structural cage.

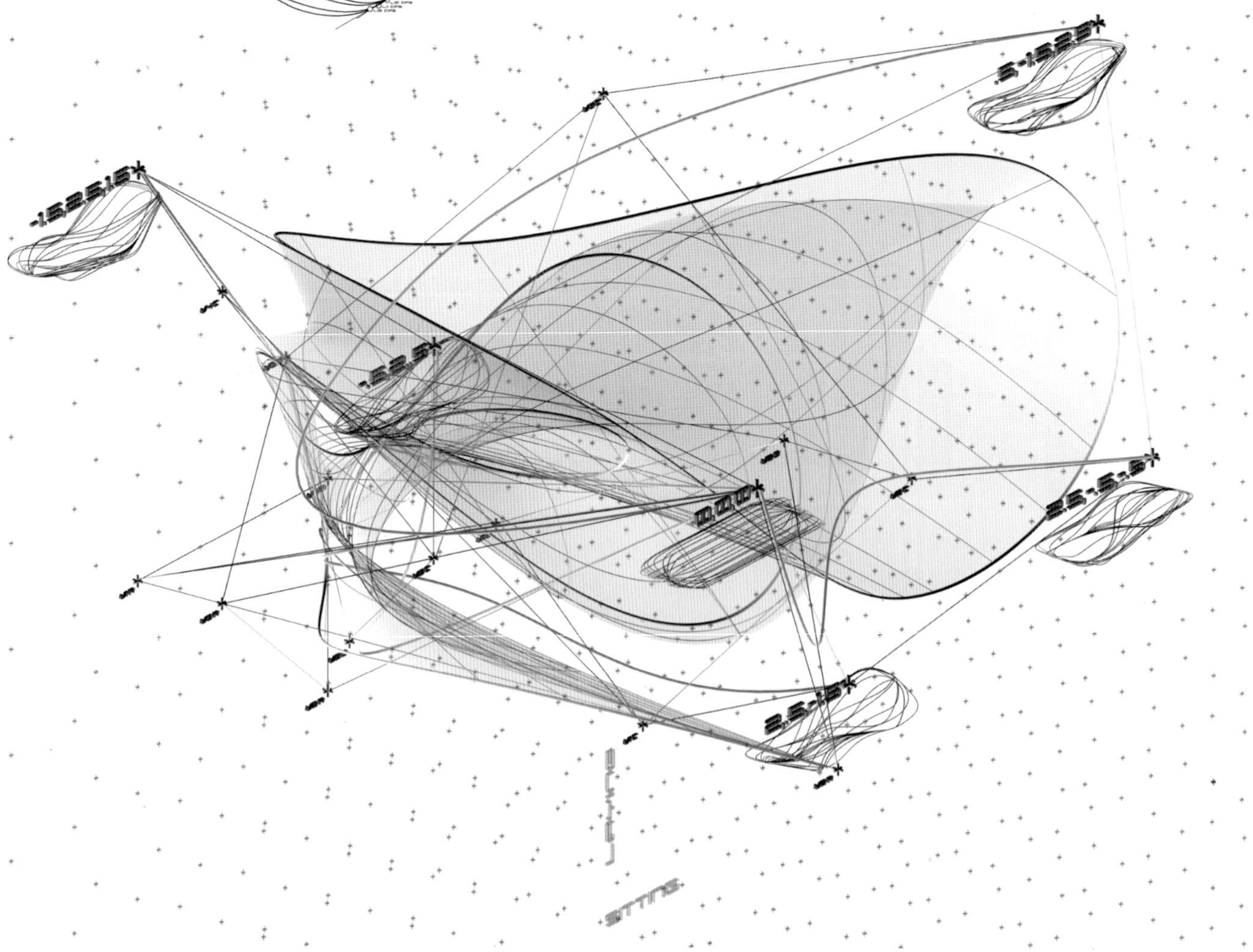

13.11
Catalytic furnishings: performance variation diagram.

We realized that if through cross-affiliation no single body position was integrated, we would have the greatest success. The body would be inflected in positions it was not accustomed to due to interactivity with the surface. Multiple users on the same surface might produce interesting or even compromising interactions between themselves, and the furnishing would engender different social arrangements depending on its location (figure 13.12).

Through the design process, we limited the number of variations (figure 13.13). This informed the design of the FlexiNurb machine and the spline structure. The forms that are produced are not referential to existing furniture types, but they might be considered to be loosely related to them. Each variation is possible within the streamlined manufacturing process we have devised. A standard thickness of material, which has been designed for the worst-case structural scenario, is placed in the mold. The

13.12 Catalytic Furnishings: scenarios of deployment.

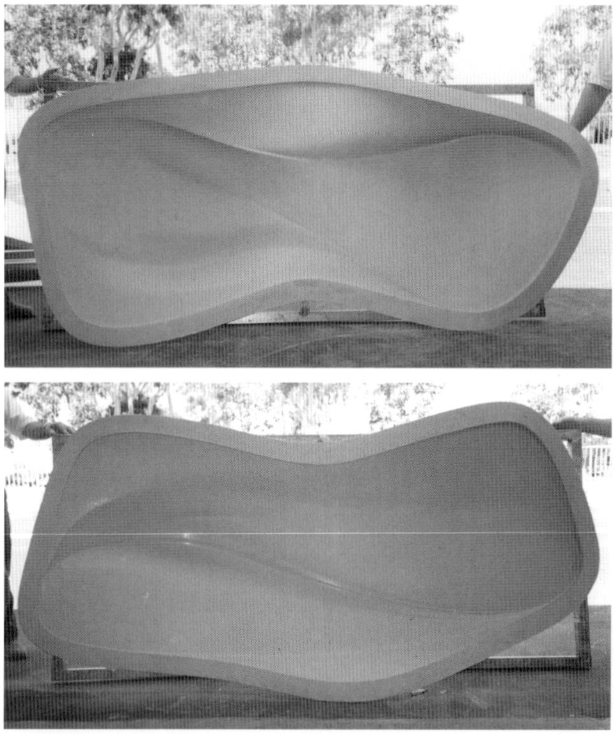

13.13
Catalytic Furnishings: two possible variations within the flexible mold system during the manufacturing process.

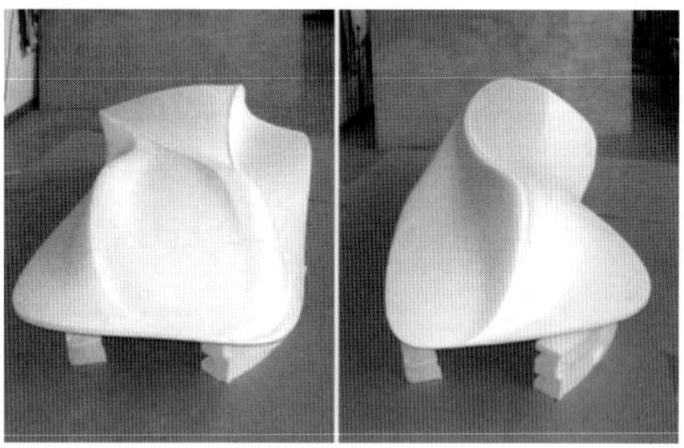

13.14
Catalytic Furnishings: the form of one possible variation made of fiberglass registers a hybridization of several lifestyles.

material is redistributed in the mold, producing a variable cross-sectional thickness. The thickness is not determined by structural efficiency, but by the effects that are desired (figure 13.14).

All possible variations are structured using fiberglass monocoque shells and are available in a wide range of finishes (figures 13.15 and 13.16). In one variation, the fiberglass thickens to allow for a gel insert, which changes the quality of the surface. Gels of varying thickness provide further gradients of softness. In another variation, we use pressure-sensitive foam that responds to body weight and allows for a different type of variation. Performative iteration is thus operative at a variety of scales, inflecting the negotiation between user and surface, and cross-profiling these inflections to different extents and in ways that produce the most emergent effects.

The flexible nurb system (FlexiNurb) responds quantitatively to different qualitative inputs (figure 13.11). One of the goals of this innovation is to standardize and automate technological processes and convert them into techniques so that they can be performed with a minimum of effort and provide stable platforms from which further innovations can be launched. The FlexiNurb process and the FlexiMold itself evolve simultaneously, and the production method becomes integral to the design of each component. Design and its implementation are no longer separated from the conceptualization process and can be thought of as one continuous process of performativity.

Moving away from efficiency and optimization towards a model for design based on performativity takes full advantage of techniques and their ability to influence behaviors and cultural endeavors. This not

13.15–16
One variation of Catalytic Furnishings being exhibited at Artists Space in New York.

only reveals greater interrelational potential between technological, material and cultural lineages, but profoundly increases their capacity to generate new effects.

The scope and significance of such change is potentially enormous. Technological choices give us a way to bridge the gap between the technical and the cultural, immersing one within the other. The feedback that is developed through this immersion creates a platform for innovation; the techniques that people generate through their use of technology not only exert pressure on technical refinement but enfold those refinements within culture. Technological choices define a world within which specific alternatives of uses can emerge, and they define a subject who chooses among those alternatives. Fundamental technological change is thus self-referential and temporal. In the making of the world through technology, we simultaneously enact great cultural change. We need to operate within technology, developing technological practices that become part of technology as opposed to applying it to whatever we are designing. Technological designing is after all ontological designing.

NOTES

1 John McKenzie, *Perform or Else*, New York: Routledge, 2001, p. 6.

2 Andrew Feenberg in *Questioning Technology* (London: Routledge, 1999) describes how the invariant elements of the constitution of the technical subject and object are modified as socially-specific contextualizing variables in the course of the realization of concrete technical actors, devices and systems. Thus technologies are not merely efficient devices, or efficiency-oriented practices, but include contexts as these are embodied in design and social insertion.

3 Larry A. Hickman, *Philosophical Tools for Technological Culture: Putting Pragmatism to Work*, Bloomington, IN: Indiana University Press, 2001.

4 A lineage is the evolutionary path demarcated by a single or combination of cultural entities, through time, as the result of replication.

5 Stephen Jay Gould, *Bully for Brontosaurus*, New York: Norton, 1991, p. 65.

6 Georges Ifrah, *The Universal History of Computing*, London: Wiley, 2000, p. 239.

7 Jeffrey Kipnis, *Mood River*, Columbus, OH: Wexner Center for the Arts, 2002, p. 53.

14

COMPUTING THE PERFORMATIVE

BRANKO KOLAREVIC

14

COMPUTING THE PERFORMATIVE

BRANKO KOLAREVIC

14.1
The Dynaform BMW Pavilion at the IAA'01 Auto Show in Frankfurt, Germany (2000–01), architects Bernhard Franken and ABB Architekten.

In avant-garde contemporary architectural design, various digital generative and production processes are opening up new territories for conceptual, formal and tectonic exploration, articulating an architectural morphology focused on the emergent and adaptive properties of form.[1] In a radical departure from centuries-old traditions and norms of architectural design, digitally-generated forms are not designed or drawn as the conventional understanding of these terms would have it, but they are calculated by the chosen generative computational method. Instead of working on a *parti*, the designer constructs a *generative system* of formal production, controls its behavior over time, and selects forms that emerge from its operation. The emphasis shifts from the "making of form" to the "finding of form," which various digitally-based generative techniques seem to bring about intentionally.

The new, speculative design work of the digital avant-garde, enabled by time-based modeling techniques, is provoking an interesting debate about the possibilities and challenges of the digital generation of form (i.e. the *digital morphogenesis*).[2] There is an aspiration to manifest formally the invisible dynamic processes that are shaping the physical context of architecture (figure 14.1), which, in turn, are driven by the socio-economic and cultural forces within a larger context. According to Greg Lynn, "the context of design becomes an active abstract space that directs from within a current of forces that can be stored as information in the shape of the form."[3] Formal complexity is often intentionally sought out, and this morphological intentionality is what motivates the processes of construction, operation and selection.

This dynamic, time-driven shift in conceptualization techniques, however, should not be limited to the issues of representation, i.e. formal appearance, only. While we now have the means to visualize the dynamic forces that affect architecture by introducing the dimension of time into the processes of conceptualization, we can begin to qualify their effects and, in the case of certain technical aspects, begin to quantify them too. There is a range of digital analytical tools that can help designers assess certain *performative* aspects of their projects, but none of them provide dynamic generative capabilities yet.

PERFORMANCE-BASED DESIGN

The aesthetics of many projects of the digital avant-garde, however, are often sidetracking the critical discourse into the more immediate territory of formal expression and away from more fundamental possibilities that are opening up. Such possibilities include the emergence of *performance-based design,* in which building performance becomes a guiding design principle, considered on a par with or above form-making.

The current interest in building performance as a design paradigm is largely due to the emergence of sustainability as a defining socio-economic issue and to the recent developments in technology and cultural theory. Within such an expansive context, building performance can be defined very broadly, across multiple realms, from financial, spatial, social and cultural to purely technical (structural, thermal, acoustical, etc.). The issues of performance (in all its multiple manifestations) are considered not in isolation or in some kind of linear progression but *simultaneously,* and are engaged early on in the conceptual stages of the project, by relying on close collaboration between the many parties involved in the design of a building. In such a highly "networked" design context, digital quantitative and qualitative performance-based simulations are used as a technological foundation for a comprehensive new approach to the design of the built environment.

It is important to note that performance-based design should not be seen as simply a way of devising a set of practical solutions to a set of largely practical problems, i.e. it should not be reduced to some kind of neo-functionalist approach to architecture. The emphasis shifts to the processes of form generation based on performative strategies of design that are grounded, at one end, in intangibilities such as cultural performance and, at the other, in quantifiable and qualifiable performative aspects of building design, such as structure, acoustics or environmental design. Determining the different performative aspects in a particular project and reconciling often conflicting performance goals in a creative and effective way are some of the key challenges in performance-based design.

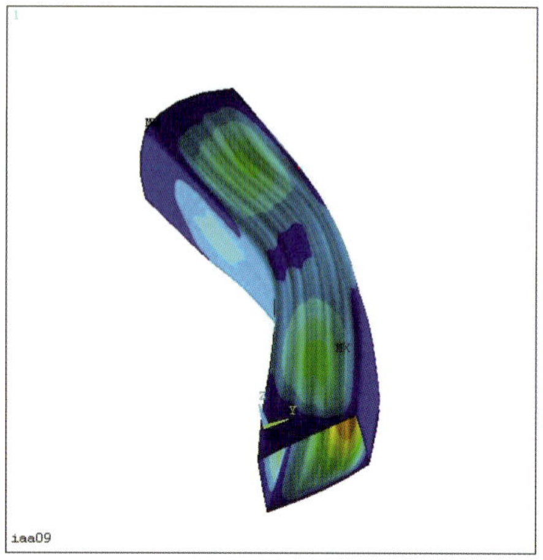

14.2
Finite-element analysis (FEA) stress analyses of the Dynaform BMW Pavilion for the 2001 Auto Show in Frankfurt, Germany, by Bollinger + Grohman Consulting Engineers, architects Bernhard Franken and ABB Architekten.

14.4
The CFD analysis of wind flows for Project ZED in London (1995) by Arup, architect Future Systems.

14.3
The FEA analysis of stresses for the Swiss Re building, London (1997–2004), by Arup, architect Foster and Partners.

CALCULATING PERFORMANCE THEN

The performative design thinking, framed by a broadly defined performance agenda and supported by a range of digital performance analysis and simulation tools, as outlined briefly above, was envisioned decades ago. Back in the late 1960s and early 1970s, a group of researchers led by Thomas Maver at ABACUS (Architecture and Building Aids Computer Unit Strathclyde) at the University of Strathclyde's Department of Architecture and Building Science, proposed that the building design be directly driven and actively supported by a range of integrated "performance appraisal aids" running on computer systems.[4]

Digital building performance "appraisal aids" and performance-based design were at the center of computer-aided building design research for more than three decades — many of the essential concepts and techniques were pioneered in the late 1960s and early 1970s. For example, the first use of computer graphics for building appraisal was in 1966, the first integrated package for building performance appraisal appeared in 1972, the first computer-generated perspective drawings appeared in 1973, etc.[5] The 1970s resulted in the "generation of a battery of computer aids for providing the designer with evaluative feedback on his design proposals," enabling architects to "obtain highly accurate predictions of such building performance measures as heat loss, daylight contours, shadow projections and acoustic performance."[6]

One of the first digital performance analysis tools to emerge was PACE (Package for Architectural Computer Evaluation), developed at ABACUS and introduced in 1970 as a "computer-aided appraisal facility for use at strategic stages in architectural design," which, unlike many of the efforts at the time, aimed "not on optimization of a single parameter but on production of a comprehensive and integrated set of appraisal measures."[7] PACE was written in FORTRAN and run on a time-sharing system; the "conversational interaction" was through a teletypewriter terminal. The program measured costs, "spatial," environmental and "activity" performance. The "spatial performance" component measured site utilization (plot ratio) and plan and mass compactness. Computing the environmental performance resulted in "plant sizes which [would] give adequate environmental conditions," while taking into account the heat gain and loss. The "activity performance" module measured "the degree to which the relationships input under activity information are satisfied by the proposed scheme."

The program would instruct the designer how to change geometrical or constructional information, i.e. how to modify the design concept to improve performance and then submit the modified design for "re-appraisal." In the end, the "repetitive man/machine interaction" would lead to "convergence of an 'optimum' design solution." A particularly interesting aspect of the program was its built-in capacity to "learn:" if the designer was satisfied with the scheme, the program would update the stored mean values used in assessments.[8]

As is often the case with visionary ideas, much of the early work in digitally-driven performance-based design was far ahead of its time both conceptually and technologically. But its time has now come, as performance-based design is slowly but steadily coming to the forefront of architectural discourse.

14.5
An early computer rendering of the structural system for Kunsthaus Graz, Austria (2000–03), architects Peter Cook and Colin Fournier (spacelab.uk).

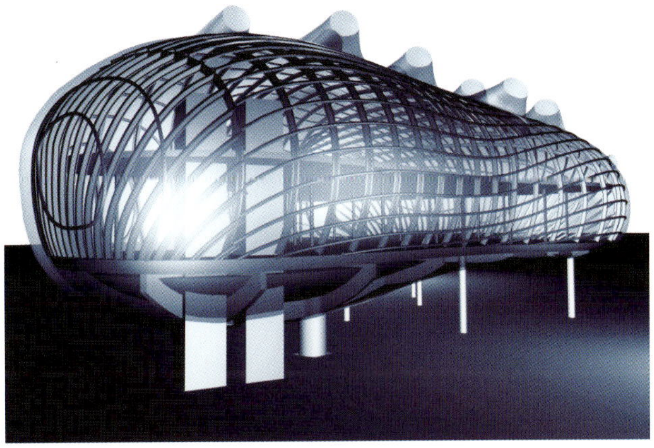

SIMULATING PERFORMANCE NOW

Today, digital quantitative and qualitative performance-based simulation represents the technological foundation of the emerging performative architecture described earlier. Analytical computational techniques based on the finite-element method (FEM), in which the geometric model is divided into small, interconnected mesh elements, are used to accurately perform structural, energy and fluid dynamics analyses for buildings of any formal complexity. These quantitative evaluations of specific design propositions can be qualitatively assessed today thanks to improvements in graphic output and visualization techniques (figures 14.2–14.6). By superposing various analytical evaluations, design alternatives could be compared with relative simplicity to select a solution that offers desired performance.

14.6
The acoustical analysis of the debating chamber in the City Hall, London (1998–2002) by Arup, architect Foster and Partners.

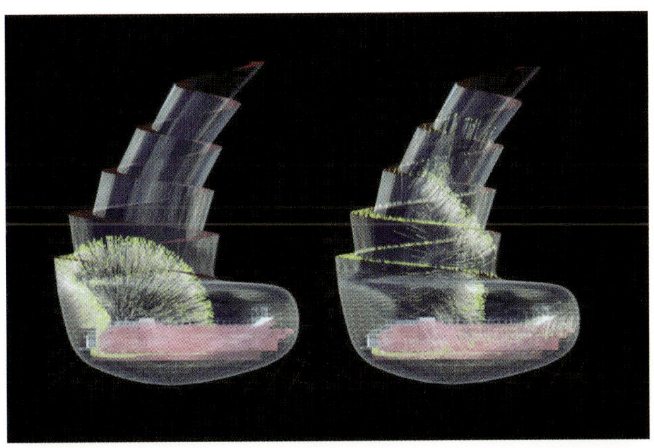

14.7
Gaussian analysis, Experience Music Project, Seattle (1999–2000), architect Gehry Partners.

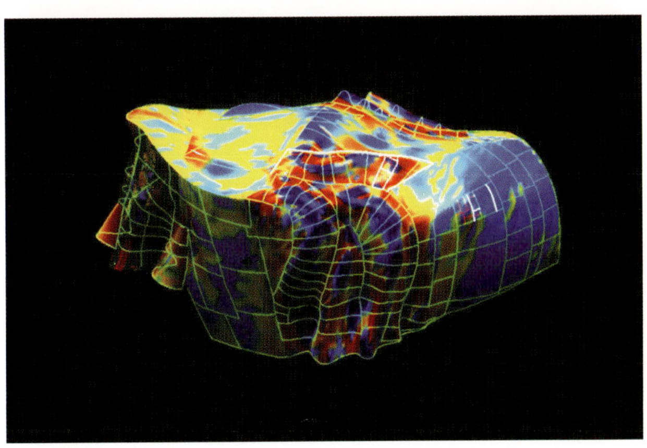

Future Systems, a design firm from London, used computational fluid dynamics (CFD) analysis in a particularly interesting fashion in its Project ZED, the design of a multiple-use building in London (1995; figure 14.4). The building was meant to be self-sufficient in terms of its energy needs by incorporating photovoltaic cells in the louvers and a giant wind turbine placed in a huge hole in its center. The curved form of the façade was thus designed to minimize the impact of the wind at the building's perimeter and to channel it towards the turbine at the center. The CFD analysis was essential in improving the aerodynamic performance of the building envelope.

The original blobby shape of Peter Cook and Colin Fournier's competition winning entry for the Kunsthaus Graz, Austria (figure 14.5), was altered somewhat after the digital structural analysis by consulting engineers Bollinger + Grohmann from Frankfurt revealed that its structural performance could be improved with minor adjustments in the overall form, by extracting the isoparametric curves for the envelope definition not from the underlying NURBS geometry but from the structural analysis. Likewise, Foster and Partners' design for the main chamber of the London City Hall (figure 14.6) had to undergo several significant changes after engineers from Arup analyzed its acoustical performance using in-house developed acoustic wave propagation simulation software.

In Gehry's office, Gaussian analysis is used to determine the extent of curvature of different areas on the surface of the building (figure 14.7). That way the designers can quickly assess the material performance, i.e. whether the material can be curved as intended, as there are limits to how much a particular material with a particular thickness can be deformed. More importantly, the curvature analysis provides quick, visual feedback about the overall cost of the building's "skin," as doubly-curved areas (shown in red) are much more expensive to manufacture than the single-curved sections (shown in green and blue tones).

As these examples demonstrate, the feedback provided by visualization techniques in the current building performance simulation software can be very effective in design development. The software, however, operates at the systemic level in the same *passive* fashion as two or three decades ago. "Computer-aided appraisal" now and back in 1980, as described by Thomas Maver, has consisted of four main elements: representation, measurement, evaluation and modification:

> The designer generates a design hypothesis which is input into the computer (representation); the computer software models the behaviour of the hypothesized design and outputs measures of cost and performance on a number of relevant criteria (measurements); the designer (perhaps in conjunction with the client body) exercises his (or their) value judgement (evaluation) and decides on appropriate changes to the design hypothesis (modification).[9]

As noted by Maver, "if the representation and measurement modules of the design system can be set up and made available, the processes of evaluation and modification take place dynamically within the design activity as determinants of, and in response to, the pattern of explorative search," which is a fairly accurate description of how performance analysis ("appraisal") software is being used today.

CHALLENGES

Designing buildings that perform (i.e. "which work — economically, socially and technically") is a central challenge for architects, as observed by Thomas Maver back in 1988.[10] He called for the development of "software tools for the evaluation of the technical issues which are relevant at the conceptual stages, as opposed to the detailed stages, of design decision-making."[11]

The challenges of developing such software, however, are far from trivial. Most of the commercially available building performance simulation software, whether for structural, lighting, acoustical, thermal or air-flow analysis, requires high-resolution, i.e. detailed, modeling, which means that it is rarely used in conceptual design development. This shortcoming, and the lack of usable "low-resolution" tools, is further compounded by the expected degree of the user's domain knowledge and skills. Another frequently encountered problem is that certain performance aspects can be analyzed in one environment while other performative analyses must be performed in some other software, often resulting in substantial and redundant remodeling. Providing a certain degree of *representational integration* across a range of "low-resolution" performance simulation tools is a necessary step for their more effective use in conceptual design.

Assuming that analytical and representational integration can be achieved, and that intuitive "low-resolution" performance simulation tools can be developed, additional challenges are presented by the need for *active* design space exploration. Instead of being used in a passive, "after-the-fact" fashion, i.e. after the building form has already been articulated, as is currently the case, analytical computation could be used to actively shape the buildings in a dynamic fashion, in a way similar to how animation software is used in contemporary architecture.[12] In other words, the performance assessment has to be *generative* and not only *evaluative*. For that to happen, however, a fundamental rethinking of how the digital performance simulation tools are conceptualized is required.

Ulrich Flemming and Ardeshir Mahdavi argued in 1993 for the close "coupling" of form generation and performance evaluation for use in conceptual design.[13] Mahdavi developed an "open" simulation environment called SEMPER, with a "multidirectional" approach to simulation-based performance evaluation.[14] According to Mahdavi, SEMPER provides comprehensive performance modeling based on first principles, "seamless and dynamic communication between the simulation models and an object-oriented space-based design environment using the structural homology of various domain representations," and bi-directional inference through "preference-based performance-to-design mapping technology."

PERFORMANCE-BASED GENERATIVE DESIGN

As Kristina Shea observed, "generating new forms while also having instantaneous feedback on their performance from different perspectives (space usage, structural, thermal, lighting, fabrication, etc.) would not only spark the imagination in terms of deriving new forms, but guide it towards forms that reflect rather than contradict real design constraints."[15] As a structural engineer, she cites the form-finding techniques used in the design of tensile membrane structures (pioneered by Frei Otto) as the nearest example of performance-driven architectural form generation, in which the form of the membrane is dynamically affected by changing the forces that act on the model. She notes that the form-finding techniques in structural engineering are generally limited to either pure tensile or pure compression structures, and she promotes the need for developing digital tools that can generate mixed-mode structural forms.[16]

14.8
Canopy design developed using *eifForm* for the courtyard of the Academie van Bouwkunst in Amsterdam (2002), designed by Neal Leach, Spela Videcnik (OFIS Architects), Jaroen van Mechelen and Kristina Shea.

14.9
eifForm: progressive generation of the canopy design.

According to Kristina Shea, a generative approach to structural design requires a design representation of form *and* structure that encodes not only (parametric) geometry but also a design *topology* based on the connectivity of primitives.[17] The experimental software she developed, called *eifForm*, is based on a structural shape grammar that can generate design topology and geometry, enabling the transformation of form while *simultaneously* maintaining a meaningful structural system. Primitives and their connectivity are added, removed and modified with a built-in *randomness* in design generation, directed by a *non-deterministic, non-monotonic* search algorithm based on an optimization technique called "simulated annealing," analogous to the "crystallization processes in the treatment of metals."[18] The software develops the overall form of a structure dynamically, in a time-based fashion, "by repeatedly modifying an initial design with the aim of improving a predefined measure of performance, which can take into account many different factors, such as structural efficiency, economy of materials, member uniformity and even aesthetics, while at the same time attempting to satisfy structural feasibility constrains." The end product is a triangulated pattern of individually-sized structural elements and joints (figures 14.8 and 14.9).

In a similar vein, I have proposed in a recent paper[19] the development of generative tools based on performance evaluation in which, for example, an already structured building topology, with a generic form, could be subjected to dynamic, metamorphic transformation resulting from the computation of performance targets set at the outset. Such a dynamic range of performative possibilities would contain at its one end an unoptimized solution and at the other an optimized condition (if it is computable), which might not be an acceptable proposition from an aesthetic or some other point of view. In that case, a suboptimal solution could be selected from the in-between performative range, one that could potentially satisfy other non-quantifiable performative criteria.

This new kind of analytical software will preserve the topology of the proposed schematic design but will alter the geometry in response to optimizing a particular performance criteria (acoustic, thermal, etc.). For example, if there is a particular geometric configuration comprised of polygonal surfaces, the number of faces, edges and vertices would remain unchanged (i.e. the topology does not change), but the shapes (i.e. the geometry) will be adjusted (and some limits could be imposed in certain areas). The process of change could be animated, i.e. from the given condition to the optimal condition, with the assumption that the designer could find one of the in-between conditions interesting and worth pursuing, even though it may not be the most optimal solution (figure 14.10).

In this scenario, the designer becomes an "editor" of the morphogenetic potentiality of the designed system, where the choice of emergent forms is driven largely by the project's quantifiable performance objectives and the designer's aesthetic and plastic sensibilities. The capacity to generate "new" designs becomes highly dependent on the designer's perceptual and cognitive abilities, as continuous, dynamic processes ground the emergent form, i.e. its discovery, in qualitative cognition. Even though the technological context of design is thoroughly externalized, its arresting capacity remains internalized. The generative role of the proposed digital techniques is accomplished through the designer's simultaneous interpretation and manipulation of a computational construct (topological configuration subjected to particular performance optimizations) in a complex discourse that is continuously reconstituting itself — a "self-reflexive" discourse in which graphics actively shape the designer's thinking process.

CONCLUSION

In conclusion, the new "performative" approach to design requires, at a purely instrumental level, yet-to-be-made digital design tools that can provide dynamic processes of formation based on specific performative aspects of design. There is currently an abundance of digital analytical tools that can help designers assess certain performative aspects of their projects *post-facto*, i.e. after an initial design is developed, but none of them provide dynamic generative capabilities that could open up new territories for conceptual exploration in architectural design. More importantly, the emergence of performance-based generative design tools would lead to new *synergies* between architecture and engineering in a collaborative quest to produce unimaginable built forms that are *multiply performative*.

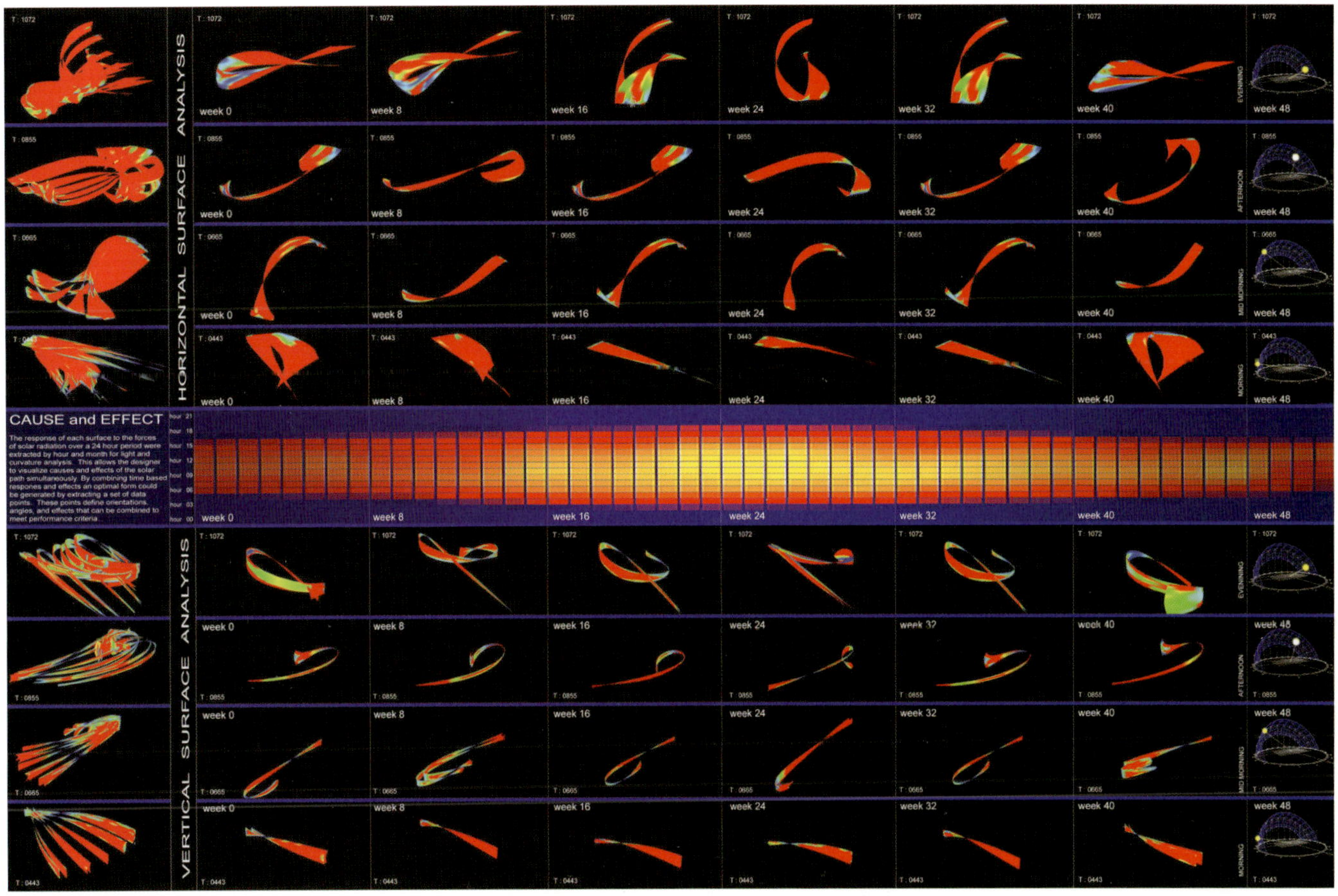

14.10
An analysis of surface curvature across a range of formal alternatives extrapolated from a computer animation by Matthew Herman (graduate student at the University of Pennsylvania).

NOTES

1 Branko Kolarevic, Chapter 2 "Digital Morphogenesis" in B. Kolarevic (ed.), *Architecture in the Digital Age: Design and Manufacturing*, London: Spon Press, 2003, pp. 11–28.

2 Ibid.

3 Greg Lynn, *Animate Form*, New York: Princeton Architectural Press, 1999.

4 Thomas W. Maver, "PACE 1: Computer Aided Design Appraisal" in *Architects Journal*, July, 1971, pp. 207–214.

5 Thomas W. Maver, "Predicting the Past, Remembering the Future" in *Proceedings of the SIGraDi 2002 Conference*, Caracas, Venezuela: SIGraDi, 2002, pp. 2–3.

6 Nigel Cross and Thomas W. Maver, "Computer Aids for Design Participation" in *Architectural Design*, 53(5), 1973, p. 274.

7 Maver, "PACE 1," op. cit.

8 The program also offered eight perspective views of the scheme, which were drawn on a "graph plotter" driven by the paper tape produced by the program. That was a "revolutionary" technological development back in 1970s!

9 Thomas W. Maver, "Appraisal in Design" in *Design Studies*, 1(3), 1980, pp. 160–165.

10 Thomas W. Maver, "Software Tools for the Technical Evaluation of Design Alternatives" in *Proceedings of CAAD Futures '87*, Eindhoven, Netherlands, 1988, pp. 47–58.

11 *Ibid.*

12 Kolarevic, op.cit.

13 Ulrich Flemming and Ardeshir Mahdavi, "Simultaneous Form Generation and Performance Evaluation: A 'Two-Way' Inference Approach" in *Proceedings of CAAD Futures '93*, Pittsburgh, USA, 1993, pp. 161–173.

14 A. Mahdavi, P. Mathew, S. Lee, R. Brahme, S. Kumar, G. Liu, R. Ries, and N. H. Wong, "On the Structure and Elements of SEMPER, Design Computation: Collaboration, Reasoning, Pedagogy" in *ACADIA 1996 Conference Proceedings*, Tucson, USA, 1996, pp. 71–84.

15 Kristina Shea, "Directed Randomness" in N. Leach, D. Turnbull and C. Williams (eds), *Digital Tectonics*, London: Wiley-Academy, 2004, pp. 89–101.

16 Ibid, p. 89.

17 Ibid, p. 93.

18 Ibid. Simulating annealing is described by Shea as "a stochastic optimisation technique that tests a batch of semi-random changes generated by the structural shape grammar, measures their performance and then chooses one that is near the best. The amount of deviation from the best is gradually reduced throughout the process, but not necessarily from one design to the next."

19 Branko Kolarevic, "Computing the Performative in Architecture" in W. Dokonal (ed.), *Digital Design, Proceedings of the ECAADE 2003 Conference,* Graz, Austria, 2003.

15

TOWARDS THE PERFORMATIVE IN ARCHITECTURE

BRANKO KOLAREVIC

15

TOWARDS THE PERFORMATIVE IN ARCHITECTURE

BRANKO KOLAREVIC

We might distinguish between two kinds of spatial disposition, effective and affective. In the first, one tries to insert movements, figures, stories, activities into some larger organization that predates and survives them; the second, by contrast, seeks to release figures or movements from any such organization, allowing them to go off on unexpected paths or relate to one another in undetermined ways.

John Rajchman[1]

In the late 1950s, performance emerged in humanities — in linguistics and cultural anthropology in particular — and in other research fields as a fundamental concept of wide impact. It shifted the perception of culture as a static collection of artifacts to a web of interactions, a dynamic network of intertwined, multilayered processes that contest fixity of form, structure, value or meaning. Social and cultural phenomena were seen as being constituted, shaped and transformed by continuous, temporal processes defined by fluidity and mediation; thus a *performative* approach to contemporary culture emerged.

As a paradigm in architecture, performance can be understood in those terms as well; its origins can be also traced to the social, technological and cultural milieu of the mid-twentieth century. The utopian designs of the architectural avant-garde of the 1960s and early 1970s, such as Archigram's "soft cities," robotic metaphors and quasi-organic urban landscapes, offered images of fantasies based on mechanics and pop culture; they have particular resonance today, as cultural identity and spatial practice are being rethought through performative acts that recode, shift and transform meanings in a true, semiotic sense.

In this spirit, performative architecture can be described as having a capacity to respond to changing social, cultural and technological conditions by perpetually reformatting itself as an *index*, as well as a mediator of (or an interface to) emerging cultural patterns.[2] Its spatial program is not singular, fixed or static, but multiple, fluid and ambiguous, driven by temporal dynamics of socio-economic, cultural and technological shifts. In performative architecture, culture, technology and space form a complex, *active* web of connections, a network of interrelated constructs that affect each other simultaneously and continually. In performative architecture, space unfolds in *indeterminate* ways, in contrast to the fixity of predetermined, programmed actions, events and effects.

The description of performative architecture given above is one of many — its paradigmatic appeal lies precisely in the multiplicity of meanings associated with the performative in architecture.[3] The increasing interest in performance as a design paradigm is largely due to the recent developments in technology and cultural theory and the emergence of sustainability as a defining socio-economic issue. Framed within such expansive context, the performative architecture can indeed be defined very broadly — its meaning spans multiple realms, from financial, spatial, social and cultural to purely technical (structural, thermal, acoustical, etc.). In other words, the performative in architecture is operative on many levels, beyond just the aesthetic or the utilitarian.

ARCHITECTURE AS PERFORMANCE

At the urban scale, architecture operates between the opposing poles of "smooth" urban space (by blending in) and urban landmarks (that stand out). Contemporary avant-garde architecture advances the latter towards architecture as performance art, which takes the urban setting as a stage on which it literally and actively *performs*.

Some of the recent projects by Lars Spuybroek (NOX), such as the D-Tower[4] in Doetinchem, the Netherlands (1998–2003), and Maison Folie[5] in Lille, France (2001–04), can literally be seen as architectural performance pieces. D-Tower is a hybrid digital and material construct (figure 15.1), which consists of a biomorphic built structure (the tower), a website and a questionnaire that form an interactive system of relationships in which "the intensive (feelings, qualities) and the extensive (space, quantities) start exchanging roles, where human action, color, money, value, feelings all become networked entities."[6] The complex surface of

15.1
D-Tower, Doetinchem, Netherlands (1998–2003), architect NOX/ Lars Spuybroek.

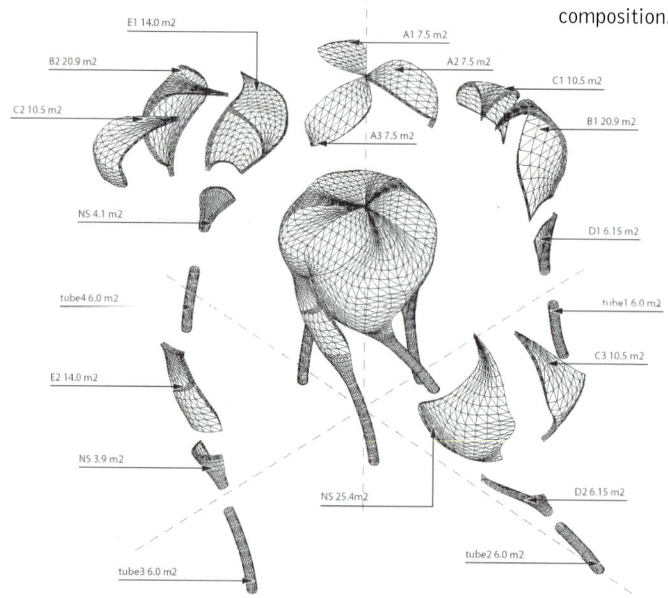

15.2
D-Tower: tectonic composition.

15.3
D-Tower: structural analysis of stresses.

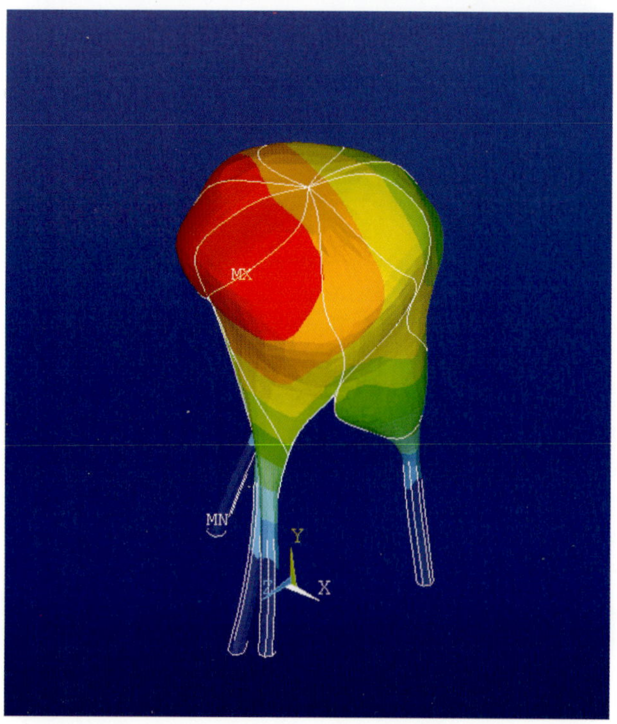

the 12 m tower is made of epoxy panels shaped over CNC (computer numeric control)-milled molds (figure 15.2). The epoxy monocoque shell is both the structure and the skin, and thus simultaneously multi-performative from the tectonic and building physics perspectives (figure 15.3). The tower changes its color depending on the prevailing emotional state of the city's residents, which is computed from responses of the city's inhabitants to an online questionnaire[7] about their daily emotions — hate, love, happiness and fear — and these are mapped into four colors (green, red, blue and yellow), with a corresponding light illuminating the biomorphic surfaces of the tower. The city's "state of mind" is also accessible through the website, which also shows the "emotional landscape" of the city's neighborhoods. So, either by looking at the tower or the corresponding website, one can tell the dominant emotion of the day.[8] The tower also features a capsule in which the city's inhabitants could leave love letters, flowers, etc. To motivate participation in this socially and culturally performative urban and architectural experiment, a monetary prize of 10,000 euros is to be awarded to the "address with highest emotions."

15.4 Maison Folie, Lille, France (2001–04), architect NOX/ Lars Spuybroek.

15.5 Kunsthaus Graz, Austria (1999–2003), architects Peter Cook and Colin Fournier (spacelab.uk).

15.6 BIX, the "communicative membrane" for Kunsthaus Graz, designers realities:united.

In Maison Folie in Lille (figure 15.4), an old textile factory that has been transformed into a new urban art center,[9] the added multi-purpose hall (a black box) features an external, partially transparent skin, whose intricate tectonic composition of metallic grilles produces varying moiré patterns as one moves along it. Spuybroek refers to this dynamic effect as a "static" movement, "an animation of the vertical tectonics of the façade, ... bending vertical lines in a complex pattern that produce a whole range of changes when walking or driving by, enhanced by the position of the sun."[10] There is also the literal movement of changing lights placed behind the metallic grille of the façade, adding another layer of intricacy to the building's urban performance.

Dynamic display of light, i.e. changing light patterns, is a primary performative dimension in Peter Cook and Colin Fournier's Kunsthaus Graz, Austria (1999–2003; figure 15.5). BIX, the light and media installation designed by realities:united from Berlin, is inserted behind the acrylic glass layer to create a "communicative membrane" — a low-resolution computer-controlled skin, a "media façade" that, through the display of signs, announcements and images, hints at the activities within the building (figure 15.6). The performative aspects of the building are all geared towards an "urban communication strategy."

The BIX light installation blurs the boundaries between the architecture and the performance medium; in the Kunsthaus Graz "the medium is the message."[11] Extending McLuhan's ideas to performative architecture,[12] one could argue that mediated, animated architectural skins have the potential to change how we relate to the built environment and, reciprocally, how the built environment relates to us, as manifested in Mark Goulthorpe's *Aegis Hyposurface* project, described below.

Movement and performance

It is often the movement of people around and through a building that gives architecture its performative capacity, as Maison Folie demonstrates. It is the experience of architecture's spatial presence and materiality — the engagement of the eye and the body — that makes architecture performative.

15.7
The Millennium Bridge in Gateshead, UK (1997–2001), architects Wilkinson Eyre Architects, engineers Gifford and Partners.

15.8
The Millennium Bridge: the bridge's arches in the tilted position.

15.9
The Milwaukee Art Museum, USA (1994–2001), architect and engineer Santiago Calatrava.

15.10
The Milwaukee Art Museum: the kinetic operation of the wing-like *brise soleil*.

In some recent projects, such as the Millennium Bridge in Gateshead, UK (1997–2001; figures 15.7 and 15.8), designed by Wilkinson Eyre Architects, and the Milwaukee Art Museum (1994–2001; figures 15.9 and 15.10), designed by Santiago Calatrava, the performative is in the kinetic effects of architecture — it is not the subject that moves but the object itself, creating an architecture of spectacle, an architecture of performance.

The Millennium Bridge in Gateshead — the "blinking eye" bridge, as it is popularly called — is the world's first rotating bridge; the entire bridge rotates around pivots on both sides of the river so that its tilt creates sufficient clearance for the ships to pass underneath (figure 15.9). The bridge's elegant arches appear to trap movement even when static; their dynamic metamorphosis has been described as resembling the slow opening of a giant eyelid — hence the "blinking eye" moniker.

For the museum building in Milwaukee, Santiago Calatrava designed a giant, movable wing-like sunscreen, a *brise soleil*, over a glass-enclosed reception hall. Made from fins ranging from 26 to 105 feet in length, the operable *brise soleil* is raised and lowered to control the amount of light (and heat) that enters into the reception area (figure 15.10). Calatrava clearly designed the operable *brise soleil* as an event, an urban performance on Milwaukee's waterfront. The performative, however, is not limited to the kinetics of the sunscreen; there are many "performances in geometry and engineering"[13] in this building, as is the case with almost all of Calatrava's projects.

In addition to kinetic effects, a building's skin can also dynamically alter its shape in response to various environmental influences, as the *Aegis Hyposurface* project by Mark Goulthorpe shows. Developed initially as a competition entry for an interactive art piece to be exhibited in the Birmingham Hippodrome Theatre foyer, the *Aegis Hyposurface* is a digitally controlled, pneumatically driven, deformable rubber membrane covered with metal shingles (figure 15.11) that can change its shape in response to electronic stimuli resulting from movement and changes in sound and light levels in its environment, or through parametrically-generated patterns. The dynamic performance of the building's skin can be either pre-programmed (determined) or in response to environmental changes (indeterminate, interactive).

15.11
Aegis Hyposurface, architect Mark Goulthorpe/ dECOi.

The Bilbao effect

In these and previously discussed projects, architecture's urban performances aim beyond the spectacle of the kinetic structures, dynamic skins and the changing light patterns. From the stakeholders' perspective (owners, municipal and regional governments, etc.), the intended performance of those buildings is primarily socio-economic; as urban landmarks, those buildings are meant to energize the urban contexts in which they are situated. By attracting the attention of local city dwellers and global cultural tourists, they are seen as the sparks of urban and economic renewal. The performances (and oftentimes forms) of these buildings become highly politicized.

This political, socio-economic and cultural performative potential of architecture is being rediscovered due, in large part, to what is nowadays called the "Bilbao effect," after the socio-economic and cultural transformation of a sleepy provincial town in northeastern Spain into a cosmopolitan cultural magnet as a result of a bold architectural and cultural strategy — the synergy of the global cultural brand of the Guggenheim Museum and the exuberance and expressiveness of Frank Gehry's architecture.[14] Not surprisingly, by reaching out for out-of-the-ordinary architectural tactics, cities increasingly expect miracles — hence, the curvaceous, light-animated forms of Kunsthaus Graz, the "blinking eye" bridge in Gateshead, and the wing-like museum in Milwaukee.

THE AESTHETICS AND ETHICS OF THE PERFORMATIVE

Admittedly, there is a considerable degree of novelty in complex, curvilinear forms (in spite of numerous precedents) pursued with fervor by the contemporary architectural avant-garde. The strong visual and formal juxtapositions created between "blobs" and "boxes" in traditional urban contexts, as is often the case, add to their "iconic" status and their perception of being exceptional and marvelous. The expressive form of the Kunsthaus Graz (figure 15.5), for example, is not accidental — its performative intent is aimed at the socio-economic: by attracting people to the area, this "Friendly Alien," as the building is curiously named by its architects, with its strange, mediated skin, will act as a development catalyst (aiming for the "Bilbao effect").

Appearance and performance

Interestingly, it is the surface — the building's skin — and its complex morphology and tectonics, and not necessarily the structure, that preoccupies the work of the contemporary (digital) avant-garde in its exploration of new formal territories enabled by the latest digital modeling software.[15] On the other hand, Santiago Calatrava appears to reject the skin in many of his projects and instead seeks to harness the expressive powers of exposed structure for its performative potential, both literally, in the engineering sense, and morphologically, for the beauty of force-driven formal articulation. Another strategy is to avoid the binary choices of skin or structure and to reunify the two by embedding or subsuming the structure into the skin, as in *semi-monocoque* and *monocoque* structures. The principal idea is to conflate the structure and the skin into one element.

This search for performance in geometry and engineering, in turn, prompted a search for different tectonics and "new" materials, such as high-temperature foams, rubbers, plastics and composites, which were, until recently, rarely used in the building industry.[16] For example, the *functionally gradient* polymer composite materials offer a promise of enclosures in which material variables can be optimized for local performance criteria, opening up entirely new material and tectonic possibilities in architecture. For example, transparency can be modulated in a single surface, and structural performance can be modulated by varying the quantity and pattern of reinforcement fibers, etc.[17]

From a historic perspective, balancing performances in geometry and material is a continuously present theme in architecture. Geometry was often imposed onto the material, as manifested by various proportioning and other ordering systems. A different approach was to let the geometry *emerge* from the material and its capacity to deal with compression and tension (i.e. the material's structural performance). Gilles Deleuze and Felix Guattari illustrate these two different approaches with a brief reference to Romanesque and Gothic architecture, where the latter represents a qualitative shift from the former, from the "static relation, form-matter" (Romanesque) to a "dynamic relation, material-forces" (Gothic).[18] As Deleuze and Guattari note, "it is the cutting of the stone that turns it into material capable of holding and coordinating forces of thrust, and of constructing higher and longer vaults."[19] The forms "are 'generated' as 'forces of thrust' (*poussées*) by the material, in a qualitative calculus of the optimum." Such "Gothic" computation of form through a material was a method, most famously, behind Antonio Gaudí's work (his inverted chain-link models) and projects by Frei Otto (the use of soap bubbles, for example). In a contemporary architectural scene, Lars Spuybroek's "analog computing" of form, accomplished through the use of threads dipped into liquids, is a direct antecedent of such a performative, materially-driven line of design thinking.[20] For many designers in the contemporary architectural avant-garde, such as Mark Goulthorpe, Lars Spuybroek, Bernhard Franken and others, the fluid synergies of form and material, appearance and performance, architecture and engineering, are intrinsically embedded into the conceptual origins of their work.

Environmental performance

Addressing the building's appearance ("how it looks") and its performance ("what it does") increasingly requires creating environmentally attuned buildings, whose physical forms are shaped by environmental performances in respect to light, heat, energy, movement or sound. There is currently an interesting gap in the aesthetics (and ethics) between form-oriented or cultural performance-oriented designers (Frank Gehry, Greg Lynn, etc.) and those whose work aims at environmental performance (Thomas Herzog, Glenn Murcutt, etc.). On the other hand, there is another group of designers — the ones whose work is neither too formalist or environmentalist (Foster, Grimshaw, Piano, Sauerbruch and Hutton, Jourda and Perraudin, etc.). The design strategies in the projects of the latter group vary considerably as they respond to different cultural and environmental contexts. In many of their projects, formal and environmental performative agendas were successfully

15.12
The Swiss Re building in London (1997–2004), architect Foster and Partners, engineer Arup.

15.13
Section drawing of the Tjibaou Cultural Center in Noumea, New Caledonia (1991–98), architect Renzo Piano, engineer Arup.

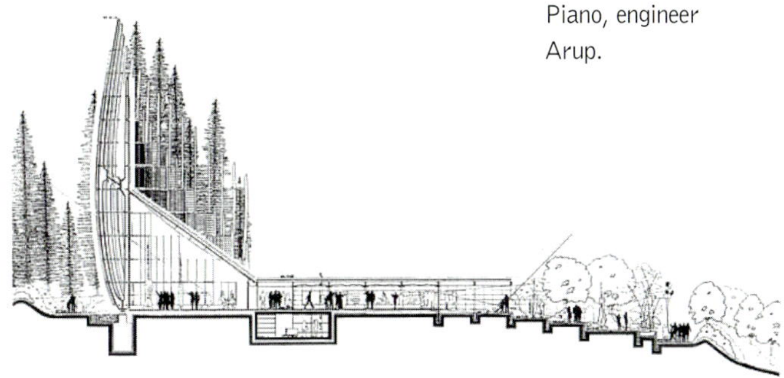

15.14
The City Hall in London (1998–2002), architect Foster and Partners, engineer Arup.

15.15
The solar diagram for the City Hall building.

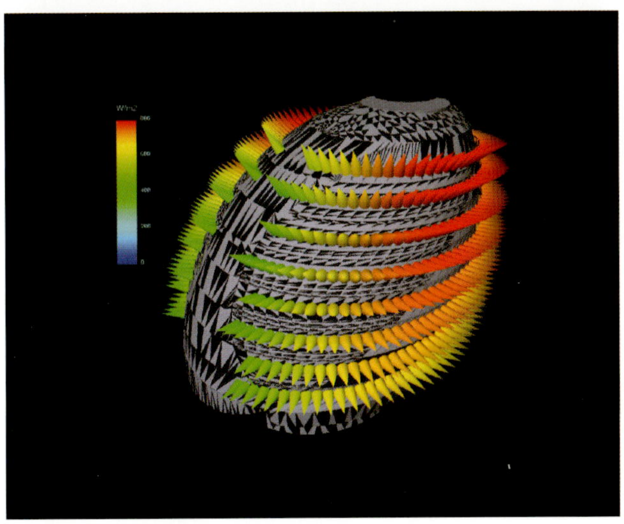

pursued in parallel. In the Swiss Re project in London (1997–2004) by Foster and Partners (figure 15.12), the design aims at maximizing the daylight and natural ventilation in order to substantially reduce (by half) the amount of energy the building needs for its operation. The spiraling form of the atria at the perimeter, which runs the entire height of the building, is designed to generate pressure differentials that greatly assist the natural flow of air. The aerodynamic, curvilinear form, besides affording a commanding, iconic presence, enables wind to flow smoothly around this high-rise building, minimizing wind loads on the structure and cladding, and enabling the use of a more efficient structure. In addition, the wind is not deflected to the ground, as is common with rectilinear buildings, helping to maintain pedestrian comfort at the base of the building.

It is interesting to note that many of the designers mentioned earlier — notably Norman Foster and Nicholas Grimshaw, once labeled High-Tech and renamed Eco-Tech by Catherine Slessor[21] — have explicitly stated their intentions to improve the environmental performance of their often highly visible buildings (figure 15.12). While one could question the methodological consistency in their projects and whether certain performative aspects, such as energy efficiency, were indeed maximized, these architects did manage to consistently push the technological envelope of environmental performance in their buildings.

An interesting example of a recent project that seems to capture the broad agenda of performative architecture, from cultural to environmental performance, is Renzo Piano's Tjibaou Cultural Center for the Kanak population of New Caledonia (1991–98; figure 15.13). The "cases" that dominate the design, and that formally reference (but do not imitate) Kanaks' huts with their cone-like shapes, were conceived with a particular cultural performance in mind. The cones of the "cases" were truncated for a more efficient environmental performance. The natural air flow within the building is then further enhanced using a system of computer-controlled louvers on the inner skin in "cases," which was designed and developed through wind-tunnel testing and computer simulations by engineers at Arup and the *Centre Scientifique et Technique du Batiment* in France.

The performative design strategies can vary considerably as they respond to different contexts. Peter Cook and Colin Fournier's Kunsthaus Graz (figure 15.5), which was discussed previously, features an expressive, biomorphic blobby form, and an acrylic glass "skin" whose primary function is to be a "communicative membrane" — a low-resolution computer-controlled skin, a "media façade." Interestingly enough, there is not a hint of environmental performance in the Kunsthaus Graz project, as if to suggest that the formal and environmental agendas are often incompatible — which cannot be farther from the truth. Foster and Partners' City Hall in London (figure 15.14; 1998–2002), imbues an iconic, biomorphic form with a logic of environmental performance that calls for such a form in the first place. (The origin of the project was purely formal — it attained its environmental logic later in the development.) The "pebble-like" form of the building in the end resulted from optimization of its energy performance by minimizing the surface area exposed to direct sunlight. The building's form is a deformed sphere, which has a 25% smaller surface area than a cube of identical volume, resulting in reduced solar heat gain and heat loss through the building's skin (figure 15.15).

Foster's performative approach to the design of the City Hall building, for example, could imply a significant shift in how "blobby" forms are perceived. The sinuous, highly curvilinear forms could become not only an expression of new aesthetics, or a particular cultural and socio-economic moment born out of the digital revolution, but also an optimal formal expression for the new ecological consciousness that calls for sustainable building.

CONCLUSIONS

Performative architecture is not a way of devising a set of practical solutions to a set of largely practical problems. It is a "meta-narrative" with universal aims that are dependent on particular performance-related aspects of each project. Determining the different performative aspects in a particular project and reconciling often conflicting performance goals in a creative and effective way are some of the key challenges in this approach to architecture.

In performative architecture, the emphasis shifts from building's appearances to processes of formation grounded in imagined performances, indeterminate patterns and dynamics of use, and poetics of spatial and temporal change. The role of architects and engineers is less to predict, pre-program or represent the building's performances than it is to instigate, embed, diversify and multiply their effects *in material* and *in time*.

The development of more performative techniques of design is essential to this task. It necessitates a shift from scenographic appearances to pragmatist imagination of how buildings work, what they do, and what actions, events and effects they might engender in time.

NOTES

1 John Rajchman, *Constructions*, Cambridge, MA: The MIT Press, 1998, p. 92.

2 Performative architecture can also be seen as a *generator* of new cultural patterns. For example, organizers of a recently held symposium on performative architecture in Delft, the Netherlands (March 11, 2004), state that "instead of describing the architectural object, performative architecture focuses on how the architectural object and its process of production perform by producing new effects that transform culture." For more details, see http://www.x-m-l.org/ and also http://www.lab-au.com/files/doc/performative_architecture.htm

3 Performance is one of the most used (oftentimes misused and abused) but least defined concepts in architecture. As can be gleaned from the chapters in this book, the ways in which performance is understood in architecture are often contradictory; the meanings associated with it are often articulated as opposites.

4 NOX (Lars Spuybroek with Pitupong Chaowakul, Chris Seung-woo Yoo and Norbert Palz) and Q. S. Serafijn, artist, and the V2_Lab (Simon de Bakker, Artem Baguinski), 1998–2003, an interactive tower, a questionnaire and a website, for the city of Doetinchem.

5 NOX (Lars Spuybroek with Florent Rougemont, Chris Seung-Woo Yoo and Kris Mun), 2001, for the city of Lille — invited competition (first prize). Model: Ouafa Messaoudi and Estelle Depaepe.

6 From the NOX Architekten website: http://www.noxarch.com

7 The questionnaire was written by the Rotterdam-based artist Q. S. Serafijn.

8 Lars Spuybroek expressed his concern that the tower could easily end up showing only one color, presumably blue (for happiness), given that Doetinchem is a Dutch city. He remarked that they may have to tweak the formula that computes the "total" emotion, so that the output is more varied. (The issue of finding appropriate "yardsticks" to measure qualitative properties that often defy quantification equally perplexes all performative domains associated with the built environment, from social dynamics to environmental comfort.)

9 The complex of buildings contains exhibition spaces, artist-in-residence homes, clubs, Turkish baths, restaurants and sound studios.

10 NOX website, http://www.noxarch.com

11 Marshall H. McLuhan, *Understanding Media: The Extensions of Man*, New York: McGraw-Hill, 1965.

12 According to McLuhan, technology effectively interferes with our senses and, in turn, affects the sensibilities of societies in which we live. That process, McLuhan argues, was and is still the cause of major cultural shifts. For more information see Eric McLuhan and Frank Zingrone (eds), *Essential McLuhan*, New York: BasicBooks, 1995.

13 Rowan Moore, "INgeniUS" in *Metropolis* magazine, June 2001.

14 According to the *Financial Times*, in the first three years since its opening in 1997, the Guggenheim Museum in Bilbao has helped to generate about $500 million in new economic activity, and about $100 million in new taxes, as reported by Witold Rybzynski in "The Bilbao Effect," *The Atlantic Monthly*, September 2002.

15 For more details, see Branko Kolarevic (ed.), "Digital Morphogenesis" in *Architecture in the Digital Age: Design and Manufacturing*, London: Spon Press, 2003, pp. 11–28.

16 For more details, see Branko Kolarevic (ed.), "Digital Production" in *Architecture in the Digital Age: Design and Manufacturing*, London: Spon Press, 2003, pp. 29–54.

17 See Johan Bettum. "Skin Deep: Polymer Composite Materials in Architecture" in Ali Rahim (ed.), *AD Profile 155: Contemporary Techniques in Architecture.* London: Wiley Academy Editions, 2002, pp. 72–76.

18 Gilles Deleuze and Felix Guattari, *A Thousand Plateaus: Capitalism and Schizophrenia*, translated by Brian Massumi, Minneapolis, MN: University of Minnesota Press, 1987, p. 364.

19 Ibid.

20 For more details, please refer to Lars Spuybroek's chapter in this volume (Chapter 12).

21 Catherine Slessor, *Eco-Tech: Sustainable Architecture and High Technology*, London: Thames and Hudson, 1998.

16

PERFORMANCE (AND PERFORMERS): IN SEARCH OF DIRECTION (AND A DIRECTOR)

PETER McCLEARY

16

PERFORMANCE (AND PERFORMERS): IN SEARCH OF DIRECTION (AND A DIRECTOR)

PETER McCLEARY

16.1
The geometry of multiple grids for services, space planning and structure — the networks are related but not coincident. Mining and Metallurgy Laboratories, Birmingham University, England (1965), architect and engineer Arup, London.

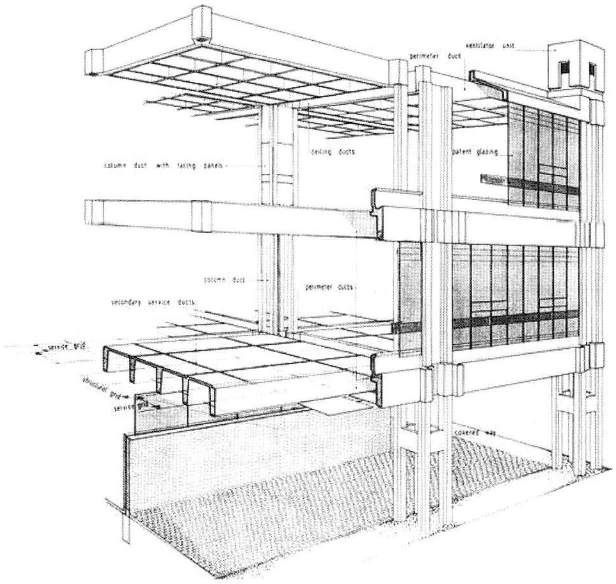

Existence is the extent of matter in space.

Descartes[1]

What stands out, that is "existence," from the homogeneity of the world is different for each of the disciplines that contribute to the production of buildings. They abstract different natural and human contexts from the total environment, attempt to satisfy different goals or performances, use different theoretical models, appropriate different materials to manifest their reality, imagine different states of equilibrium from open to closed or active to passive for their systems, engage different builders and building processes to construct their designs, generate different flow patterns or vector diagrams to transmit their particular forms of energy, and use different methods to synthesize their disparate views of the environment.

The production (design and construction) of buildings is a complex system. The present deconstruction of the system into many different subsystems of professions and crafts, theories and practices, and materials and methods, illustrates that ontogeny recapitulates phylogeny, i.e. today's design and construction processes result from their own historical development.

The difficult task is to combine the disparate perceptions of the performances to be satisfied, and to synthesize the geometry of the different flow patterns (static and dynamic) into a coherent whole. Possible future solutions might derive from Herbert Simon's avoidance of optimized subsystems in favor of "satisficing" each subsystem through negotiation and coordination,[2] the hierarchical deconstruction of the whole system, or change from a "state description" of the world as sensed to a "process description" of the world as acted upon.

If "performative architecture" puts the satisfaction of human equilibrium at the center of its purpose, a radical change needs to take place in the representation of the systems of the architectural design and building process.

Many of today's fashionable ideas were equally fashionable in the 1960s; that is, the General System Theory, cybernetics, design methods, and mathematical and physical biology. While today we revisit those ideas, in those earlier days calculations were constrained to log tables, then slide rules, and finally to electronic hand calculators. In 1961, in the structural division of Ove Arup in London, much time was spent inverting matrices (related to the geometry of the Sydney Opera House) with a hand calculator — they did not have a computer. Now that there are bigger, faster calculating devices, larger, more complex problems can be solved. In that distant past, such problems could not be examined without the use of a theory of hierarchies and methods of synthesis.

The concept of this chapter does not derive from my research at Penn in the 1970s with Robert Le Ricolais and the rhetoric of tension, or with Louis Kahn and his search for essences. Thus, there is no mention of the rheology of matter and structure and the isotropy of space. Rather, the idea is drawn from my 1960s professional experience with the Ove Arup Building Group, later named Arup Associates.

Their key concept was that "the design and construction of buildings was the task of a multi-professional team."[3] This core idea was not a mere auxiliary hypothesis to other theories of architecture and engineering, but was sufficiently profound to deflect my studies from the uniqueness of role of structure in building to that of collaboration, coordination and towards the belief that the design activity is polygamous (figure 16.1).

This chapter has three parts: a short prelude or introductory section to its fugue, which is a polyphonic composition (like Arup Associates, with their interweaving of the logic of the many professional disciplines), and it concludes with a coda on the synthesis of those logics.

PRELUDE

In 1951, ten years before I joined the Ove Arup Building Group, there was a conference and exhibition in Darmstadt on the subject of "Man and Space." It served as a commemoration of the 1901 Jugenstil exhibition on Behrens, Olbrich and others. The 1951 exhibition exhibited photographs of the work of some twentieth-century architectural masters: Taut, Loos, Le Corbusier, Gropius, Behrens and Olbrich again, Frank Lloyd Wright, and others.

On the last day for the colloquium, or academic conference, Martin Heidegger presented his essay on "Building Dwelling Thinking,"[4] and Jose Ortega y Gasset spoke on " The Myth of Man Beyond Technique."[5] Among Heidegger's hypotheses are that building has dwelling, that is, care, concern, cultivation, etc., as its goal; that dwelling and building are related as ends and means; and that to be human means to dwell. His explication uses the case study of a bridge. To an engineer, this is a revealing interpretation of the nature of a bridge.

Ortega, often at odds with Heidegger — although both were students of Edmund Husserl[6] — argues that dwelling does not precede building, and as man has not adapted to the world (i.e. he is an alien), he does not belong, he needs a new world, thus he wants to build "like a lover without a beloved;" dwelling is not given to man, rather he fabricates it — and dwelling is a privileged position or situation.

Apparently there were mixed reactions from the architects present at the colloquium, ranging from a general uneasiness to questioning the validity of the participation by philosophers.

This short prelude does not focus on this difference between the positions of Heidegger and Ortega, that is, on the primacy or priority of building and dwelling. Instead, we note some similarities between their thoughts on the subject of technology.

Not content with his contribution, in Darmstadt, to the discourse on dwelling and building, Jose Ortega y Gasset (1883–1955) continued to explore the topic in his essays, leading to "Pragmatic Fields."[7] This well-known essay was published in Spanish after his death in 1955. In it he suggests that our lives consist in the articulation of many small worlds or territories (e.g. religion, knowledge, business, art, love), and that our life is nothing but a relentless dealing with things. Properly speaking, in life there are no "things," only in scientific abstraction do things exist, and realities that have nothing to do with us are just there, by themselves, and independent from us. However, we must deal and occupy ourselves with some things that are issues, that is, something that must be done — a *faciendum* in Latin, or *pragmata* in Greek. He concludes that, therefore, we must contemplate our lives as an articulation of "pragmatic fields."

In his earlier 1939 essay, "Meditacion de la Tecnica" (or "Thoughts on Technology," or "Man the Technician"[8]), Ortega says that humans are ontological centaurs, half immersed in nature, half transcending it: the extra-natural part is a program (or project) of life and that that program is limited to a pragmatic field. If we agree, we must conclude that, despite rumors to the contrary, we live in different worlds.

Much earlier, in his 1926 "Being and Time,"[9] Martin Heidegger (1889–1976) had considered aspects of the topic of the 1951 conference. His "analysis of environmentality" discusses the spatiality of the "ready-to-hand" or the poetics of use, in which we discover the world through touch and making things, and that this activity has its own kind of knowledge.

The focus of this volume seems to be on the physics, not poetics, of building. Heidegger, or Husserl, might say the emphasis is on the "present-at-hand." He proposes that humans encounter the world through equipment, and that since they use equipment in order to do something, it has the character of closeness (or nearness), and further that this closeness of equipment has been given directionality. (See my essay on the transparency-opacity and amplification-reduction of equipment.[10])

Heidegger continues that equipment has its place and this place defines itself as the place of this equipment:

> Thus the sun whose life and warmth are in everyday use has its own places — sunrise, mid-day, sunset, midnight. The house has its sunny side and its shady side; the way it is divided up into rooms is oriented towards these, and so is the arrangement within them, according to their character as equipment. Thus we discover the spatiality of equipment. And space or geometric space has been split up into places through equipment.[11]

It is this similarity between Ortega and Heidegger, that is, the existence of worlds that we experience through mediation, that leads one to seek parallel hypotheses in natural philosophy, in addition to human, ontological or metaphysical philosophy.

There are parallels in Ernst Mach's "Space and Geometry,"[12] where he correlates the separate physiological spaces of vision, touch, smell, sound, and movement, and in the idea of Descartes that existence is the extension of matter in space.[13]

The conclusion to this prelude, drawn from Arup Associates, Husserl, Heidegger, Ortega, Mach, and Descartes, is that each of us live in a different abstracted world that is made apparent by our forms of mediation (equipment, processes and theories), and that we use not only different means but also imagine different ends.

FUGUE

Applying this conclusion to the topic of "performative architecture," we note that each time an aspect of architecture, that represents a particular abstracted world, is rendered measurable, there arises a new concomitant specialty, whose authority is legalized through licensure. What is common to all these special disciplines is that there is a continuum, real or imaginary, applicable to the whole architectural environment and to the specialized or abstracted parts. We will examine aspects of this continuum with its environmental contexts (human and natural), goals (purposes and performance), theoretical models of each discipline, the materials and equipment used, the building processes, their individual system geometries (or vector analyses), and the task of the synthesis of the separate systems.

Environmental contexts

From the human context, the architect considers relevant social structures (family, tribe, etc.), new agreements, institutions, human comfort, ergonomics, size and orientation of activities, movement of people, and socio-cultural preferences and values.

From the natural context, the structural engineer recognizes the imposition of load actions due to gravity (self-weight and applied load), wind, snow, earthquake, ground movement and temperature changes.

Other specialists, often engineers, respond to changes in heat or coolness (skin), light (eye), sound (ear), smell (nose), air quality (lungs), and the flow, in both directions, of liquids and gases (e.g. plumbing).

Apparently, they experience different worlds, or at least they are expected to examine different environments.

Performance

Each of these professional realities seeks to satisfy different ends or purposes. The architect aims to optimize a number of utility functions, that is, the product should be useful. Among those functions of what was formerly referred to as *utilitas*, is accommodation to the body (ergonomics) and its activities (space planning), or even to design spaces that "inspire" activities (Louis Kahn). Other utility functions consider human comfort and health (physiological), environmental stable equilibrium (homeostasis), safety and security, durability, and the efficiency, life-cycle costs of capital, labor and energy.

The architect must also consider issues of appearance or grace, once named *venustatis*. Here the appeal is to delight and pleasure, both at the level of individual preferences and cultural values. Sometimes the goal reaches beyond the concerns of beauty towards the realm of the sublime.

A structural engineer examines, in part, issues of strength, once named *firmitas*. Here the expectation is for stability or no collapse, and strength or no yielding. To achieve these goals with efficiency, the task is to minimize potential energy and to maximize strain energy.

The specialist in light considers general illumination and task lighting, but today rarely considers metaphysical light. To perceive the space, this engineer must illuminate the objects to place forms in relationship to each other.

The acoustics specialist considers the comfort of human hearing.

And so forth.

The different professions are responsible for different environments; they experience different worlds, and thus have different goals in mind. The problem remains to combine these different perceptions.

Theoretical models

Each specialist has a different theoretical model whose language ranges from verbal through graphical to mathematical. These theories are at different stages of development in a continuum from magic through craft (e.g. rules-of-thumb) and empiricism (e.g. analogies, codes of practice) to applied science.

Most theoretical models in the building industry are closed systems from mechanics that focus on the stationary state. Today's interest is in open or living biogenetic or biophysical systems in steady state equilibrium. In the near future we will rediscover the much fuller explication of systems that includes kinetics, dynamics and statics with its equilibria of the steady state, moving equilibria, and displacement of equilibrium. Perhaps even the principle of Le Chatelier and Alfred J. Lotka's 1924 text *Elements of Mathematical Biology*[14] will be rediscovered. There are many precursors to our present interest, in particular, and to general systems. Meanwhile, each specialty uses theories unique to its discipline.

The architect models reality with a vector analysis of size, orientation and movement, flow of activities, concepts of space, such as classicism, modernism, deconstructivism, and emerging theory of complexity. Typology and morphology offer palaeontological theories of space.

The structural engineer theorizes a mechanical world of statics and dynamics, linear and non-linear, scientific analogies, graphical methods, algebra, calculus,and matrices. In G. H. Hardy's essay, " A Mathematician's Apology,"[15] he says that matrices are beautiful because they are not useful in modeling physical problems. While with Ove Arup, I tediously inverted matrices that modeled some aspect of the geometry and structure of the Sydney Opera House. Is it possible that today there exist some beautiful, useless ideas that in time will become useful? A sort of latent utility lies in the beautiful. Of course, the balance between potential energy and strain energy is at the core of the structural engineer's thinking, as is the need for upper and lower bound solutions. Problems related to scale, or relative size, use the principle of similitude or the theory of dimensions and its understanding of the fundamental variables of mass, length and time.

The fact that many architects imagine the relationship between structure and space in terms of mass alone results in architectural theories of structural behavior that are appropriate only to heavy masonry structures. Most engineers perceive space in a world of framed structures restrained by inertia, and some can visualize the space of pre-stress and force.

The specialist on heat imagines a thermodynamic reality where the desire is for a "thermal steady state across time and thermal equilibrium across space." Theirs is a world of forms of energy, microclimate, heat transfer, band width of radiation, heat that is sensible, latent, radiant or specific, and is transmitted by radiation, conduction or convection. In addition, their measures are of resistance and conductivity, thermal lags, solar radiation, mean temperatures and shading coefficients. To their perception, the presence of the sun is more evident than the pull of gravity. Heat is often aligned with humidity and its theories of metabolism, absorptivity, emissivity and charts that are either bioclimatic or psychrometric.

Light and color have an array of theoretical models derived from optics, with its rays that are absorbed, reflected, refracted or transmitted.

Acousticians speak of sound pressure waves, frequency and wavelength, reflection, absorption, transmission, reverberation and resonance, pitch and tone.

Theories of the behavior of fire consider the importance of flame spread, flammability and toxicity.

Builders also have operative theories that deal with the operations of man-machine complexes. These theories attempt to formulate a scientific theory of action, sometimes named "praxiology."[16] Other models are drawn from operations research, decision theory and cybernetics.

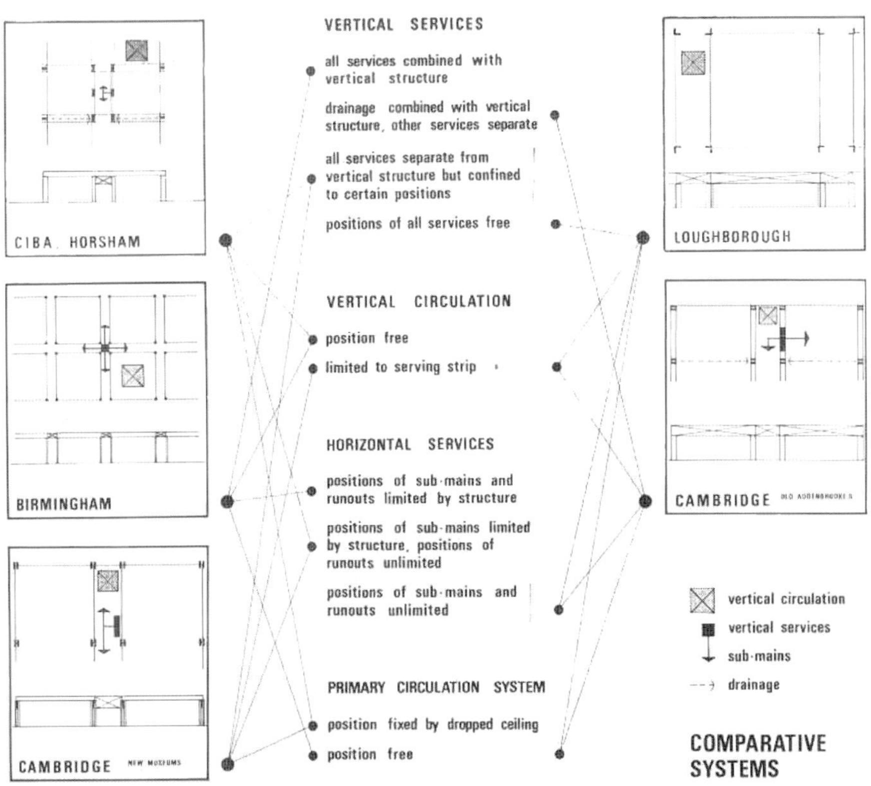

16.2
Diagram illustrating a comparison of systems devised for science buildings. Comparative systems of buildings designed by Arup Associates in Birmingham, Cambridge, Horsham and Loughborough, England; architect and engineer Arup, London.

Units of measure

Each of the theoretical constructs conjures up its unique units of measure, notwithstanding that the theory of dimensions shows that all units are functions of mass, length or time. Whereas the architect might emphasize length, area and volume, the structural engineer measures in weight, stress, strain, modulus of elasticity, inertia and so forth. The other specialists focus on the degree, thermal units per hour, joule and watt, candela, lumen, foot-candle, mass and decibel.

The many different languages, using different theories, words and units of measure, that are spoken in the design and construction of buildings can lead to babel, that is, a confusion of tongues.

Materials

To achieve its goals, each discipline uses its own specific building materials. The architect might focus on materials that relate to perception of the space. The structural engineer uses many materials from the most ancient masonry and wood through steel and reinforced or prestressed concrete to present-day carbon fiber. In the future, liquids and gases might replace the transmission of forces through solids. Similarly, HVAC, lighting, acoustical and fire engineers, all use specialized materials to accommodate their different perceptions of building performance.

Processes

During the processes of manufacture, fabrication and assembly, different subcontractors assemble each of these different materials, each using their own particular tools, processes and theories. Builders have their own mini-max notions of efficiency.

Geometries

Each profession articulates the geometry for efficient patterns, layouts or types or system geometries for vector flow, both its direction and magnitude, of the flux of the movement of people, air, stresses, waves (heat, light, sound, smell), liquids (water and sewage), and the rhythm or metric of the construction process (figure 16.2).

While some architects are more interested in solids than voids, it is more usual for them to examine the flow through the voids, and describe its dimensions (linear, planar or volumetric), direction (one or two way), its orientation to some reference datum, horizontal and vertical, the profile of the boundaries in terms of an inside surface (or *intrados*) and outside boundary (or *extrados*), concavity or convexity. Today we see more use of complex geometrical shapes that shows some knowledge of Gaussian curvature, e.g. torqued ellipses. Perhaps tomorrow we will see some use of the seven elementary catastrophes, e.g. the hyperbolic umbilic. Similarly, there is an increasing awareness of the topology that models the connectivity of spatial activities and their adjacencies. There remains the most basic understanding of flow as circulation through entrance, foyer (at rest), corridor (in motion), stairs (incline and algebra), ramps (incline and calculus), escalators (moving incline), elevators (vertical), and so forth.

The degree of isotropy, or the proportional relationships among the dimensions of space and time, and topological measures are significant measures in the geometry of architectural space.

The structural engineer is more concerned with rheology, or the flow of stresses and deformations in the solid matter. Similar to the voids of the architect, this geometry describes the solids in terms of their axes of restraint in one, two or three dimensions, and the flow or "spanning" in one, two or three directions. Various shapes are characterized as singly or doubly curved, synclastic or anticlastic. The familiar shapes are the cylinder, sphere, conic and hyperbolic sections, torus, and so forth. Rene Thom's research on "chaos theory,"[17] with its geometry of seven "elementary catastrophes," has brought interest to the shapes of the fold, cusp, swallowtail, butterfly, and three umbilic forms — hyperbolic, elliptic and parabolic. These complex shapes must face the scrutiny of stress analysis and they may or may not be as useful as minimum volume networks of evolute and involute curves.

Studies in the isotropy of geometric space and the rheology of matter show that space and matter are geometric duals, that is, "images" of each other.

The expert on heating, cooling and air conditioning studies the geometry of the flow in liquids, gases and waves. They imagine open or closed channel flow in ducts or pipes. Among their systems are two-pipe, four-pipe, closed loop, central and peripheral, and single or multi-zone. These geometries allow more systematic and systemic thinking than the more general typology of the hearth, foyer, inglenook, gazebo, porch, atrium and greenhouse. Generally, the existence and use of typologies indicates a formal logic in its infancy. Specialists in lighting and acoustics also have a set of preferred configurations that satisfy their performance criteria. Builders too have their intuitive patterns and rhythms that derive from critical path methods, programmed enquiry research techniques and more recent computer models that simulate the building process. Soon there will be a certified materials engineer who alone knows the performance of materials.

Clearly, each profession lives in a different world, has different goals, constructs different theories, uses different materials and methods of building, and imagines a different geometry.

CODA: SYNTHESIS AND THE INTERWEAVING OF COMPLEXITY

To synthesize these disparate perceptions we must look deeper than "design value analysis" and "economic evaluation methods," such as payback period, life-cycle cost, return on investment, or comparative value analysis.

The production, both design and construction, of buildings is a complex system. The present deconstruction of that system into many different subsystems of professions and crafts, theories and practices, and materials and methods, illustrates that ontogeny recapitulates phylogeny. In other words, the organization of design and construction processes is the result of their own fragmented perception and historical development.

16.3
The geometry of relationships among services, space planning, structure and partitions: Loughborough University of Technology Laboratory, Loughborough, England (1965), architect and engineer Arup, London.

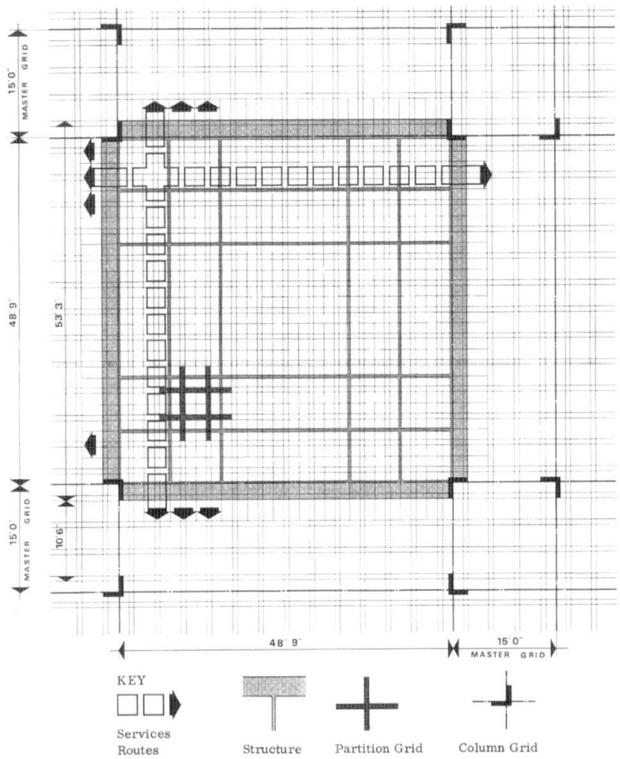

In this complex system, the flow path of each optimized subsystem interferes with the paths of the other systems (figure 16.3). In the 1970s, Philip Dowson, an architectural partner of the Arup Building Group, wrote that "as each profession operates only within prescribed limits and carefully defined boundaries of responsibility and authority, then an uncoordinated result must be expected," and that "with five of the seven professionals within an integrated design group having their training largely technically-based, it will be inevitable that design solutions will be technically biased."[18]

This suggests the need for team members from other disciplines — perhaps from human-factors, sociologists, physiologists, behavioral and perceptual psychologists, and philosophers — notwithstanding the reservations of the architects at the 1951 Darmstadt colloquium.

What methods might reconcile these seemingly contradictory possibilities of optimized subsystems? The difficult task is to make a coherent whole by interweaving the geometry of the multiple flow patterns.[18] Each specialist can describe several alternative systems in terms of their geometric patterns and concomitant materials. The team, under the leadership of the architect, negotiates the interweaving of the geometries to where all the systems are satisfied but not optimized.[19]

As stated in the introduction, Herbert Simon's "Sciences of the Artificial"[20] describes a similar method that avoids optimizing subsystems by "satisficing" each subsystem through negotiation and coordination. He also suggests benefits in a change from a "state description" of the world as sensed to a "process description" of the world as acted upon.

Cybernetics[21] also suggests a theory of hierarchies[22] to deconstruct the whole system. It proposes diagrams of immediate and ultimate effects, where "intense interaction implies spatial propinquity."[23]

This is akin to Heidegger's idea that humans "live in the space opened up by equipment."

The above deconstruction of the whole system of building is similar to a medieval scholastic system theory[24] in which they seek unity in diversity, that is, the *unitas multiplex*. The systemic relationship involves a decomposition of the whole into parts, or *explicatio*, and the recomposition, or *implicatio*. The parts are related to the whole hierarchically, or *concordantia*. And, as in engineering aesthetics the logic should be seen, or *manifestatio*, and seen clearly, or *claritas*.

A temporary conclusion is that, if "performative architecture" puts the satisfaction of human equilibrium at the center of its purpose, a radical change needs to take place in the representation of the multidisciplinary systems of the architectural design and construction process. Equally important is that research must not be limited to the different perceptions of the world, but must expand to include predictive theories that synthesize these many different perceptions.

NOTES

1 Stephen Gaukroger, *Descartes: An Intellectual Biography*, Oxford: Oxford University Press, 1997.
2 Herbert A. Simon, *The Sciences of the Artificial*, Cambridge, MA: MIT Press, 1969.
3 Philip Dowson, *Architecture and Professional Integration*, UK: Public Works Congress, 1972.
4 Martin Heidegger, *Poetry, Language, Thought: Building, Dwelling, Thinking*, New York: Harper, 1971.
5 Jose Ortega y Gasset, "The Myth of Man Beyond Technique" in *Obras Completas*, Madrid: Revista de Occidente, 1958.
6 Edmund Husserl, *Ideas: General Introduction to Pure Phenomenology*, London: Collier MacMillan, 1962.
7 Jose Ortega y Gasset, "Pragmatic Fields" in *Obras Completas*, Madrid: Revista de Occidente, 1958.
8 Jose Ortega y Gasset, *History as a System: Man the Technician*, New York: W. W. Norton, 1962.
9 Martin Heidegger, *Being and Time*, translated by J. MacQuarrie and E. Robinson, New York: Harper & Row, 1962.
10 Peter McCleary, "Some Characteristics of a New Concept of Technology" in J. M. Stein and K. F. Spreckelmeyer (eds), *Classic Readings in Architecture*, Boston, MA: McGraw-Hill, 1999, pp. 169–179.
11 Heidegger, *Being and Time*, op. cit.
12 Ernst Mach, *Space and Geometry: In the Light of Physiological, Psychological and Physical Inquiry*, Chicago, IL: The Open Court Publishing Co., 1906.
13 Rene Descartes, "Principia Philosophiae" in Charles Adam and Paul Tannery (eds) *Oeuvres de Descartes*, 13 volumes, Paris: Leopold Cerf, 1897–1913.
14 Alfred J. Lotka, *Elements of Mathematical Biology*, New York: Dover Pub., Inc., 1956.
15 G. H. Hardy, *A Mathematician's Apology*, Cambridge: Cambridge University Press, 1967.
16 Tadeusz Kotarbinski, *Praxiology: An Introduction to the Science of Efficient Action*, New York: Pergamon Press, 1965.
17 Rene Thom, *Structural Stability and Morphogenesis*, Reading: Benjamin, 1975.
18 Derek Sugden, "Planning for the Unknown" in *British Conference on Steel in Architecture*, November 1969, pp. 157–167.
19 Phllip Dowson, "Building for Science" in *Architectural Design*, April, 1967, pp. 160–170.
20 Simon, op. cit.
21 Norbert Weiner, *Cybernetics: On Control and Communication in the Animal and the Machine*. Cambridge, MA: MIT Press, 1965.
22 Howard H. Pattee, *Hierarchy Theory: The Challenge of Complex Systems*, New York: Braziller, 1973.
23 W. Ross Ashby, *Design for a Brain*, London: Chapman and Hall, 1966; W. Ross Ashby, *An Introduction to Cybernetics*, London: Methuen, 1968.
24 Erwin Panofsky, *Gothic Architecture and Scholasticism*, New York: Meridian, 1957; Maurice De Wulf, *The System of Thomas Aquinas*, New York: Dover Pub., Inc., 1959; Umberto Eco, *The Aesthetics of Thomas Aquinas*, translated by Hugh Bredin, Cambridge: Harvard University Press, 1988.

17

CONCEPTUAL PERFORMATIVITY

WILLIAM BRAHAM
HARALD KLOFT
DAVID LEATHERBARROW
ALI RAHIM
MAHADEV RAMAN
ANDREW WHALLEY
BRANKO KOLAREVIC
ALI M. MALKAWI

17

CONCEPTUAL PERFORMATIVITY

WILLIAM BRAHAM
HARALD KLOFT
DAVID LEATHERBARROW
ALI RAHIM
MAHADEV RAMAN
ANDREW WHALLEY
BRANKO KOLAREVIC
ALI M. MALKAWI

KOLAREVIC: We have heard very different versions or visions of what performativity or performance means in architecture; as we have said at the outset, such an outcome was expected. The challenge is to give some kind of overall coherence to the notion of performativity or performance in architecture.

In the two panel discussions we will examine a thematic territory that lies between what we see as two poles of performativity. At one end are the *conceptual* domain and the *performativity of the virtual*. At the other are the *operative* domain and the *performativity of the real*. We hope to situate the discussions between these two poles — the conceptual realm and the operative realm — perhaps getting a little closer to the first one in this discussion, and to the second one in the discussion that follows. We are not really constraining this discussion to the conceptual realm as, in my view, it is impossible to separate the two.

We will begin the discussion with a few questions and issues associated with the conceptual realm. To what extent do the issues of performance, broadly understood as we have heard, figure in conceptual design? Is there such a thing as performative form-making or performative space-making? What is the relationship between performance simulation and form-making, and how can one inform the other and vice versa?

So, the first question is whether there is such a thing as performative space-making or performative form-making?

WHALLEY: I saw David Leatherbarrow's definition of performance in architecture as having two halves, and I was actually trying to work out in which of the two our work would be located. Undoubtedly, our early work sits in the first half, which is highly programmed, highly functioning, program-specific spaces. As our work has evolved, it was clearly moving and blending towards the second.

In the early days we were really quite obsessed with function, to the exclusion of all other things. For instance, the early factories were entirely about designing a skin and that was it. To some degree such disposition also had to do with the type of architecture we were given at the time, which was fairly simple architecture, low budget, so the outcome was a one-liner.

As projects became more sophisticated and complex, our work started to take on broader aspects, in addition to just being purely functional. In other words, at the end of the day we actually stepped back and asked: but is it beautiful? Is there a materiality to it? How will it weather? How will it look in twenty years' time? How will it sit in this environment? How will people relate to it? Is it the right tactile material on the inside?

There is a whole series of filters as well as the pure programmatic. To some degree, high-quality architecture has to take both into account. Architecture has to be programmatic, it has to be functional, it has to deliver what people want it to do. One cannot just exclude all of that.

To some degree, David Leatherbarrow was showing architecture that was just architecture in its own right and then took on a programmatic quality, in addition to its sheer beauty and tactile quality. For me, architecture has to move towards that programmatic quality ...

KOLAREVIC: It has to perform?

WHALLEY: It has to perform, but at the same time it has to also take on these other levels. One cannot just work at one level.

KOLAREVIC: So, would you say that Grimshaw is engaged in performative space-making?

WHALLEY: Well, yes. As I explained, our work is process-driven; it is not stylistically predictable — it is very much driven by the individual program of a particular project. Through that process, we take on many other architectural issues, such as the issues that have to do with the art of architecture. When we used the water wall to cool the British Pavilion building, we designed a functioning system to demonstrate how it could work. But the project is actually about the senses, about creating the feeling of an oasis of a cool space in a hot climate, of stepping through the sound of water, stepping through into a pool, gliding over that pool and seeing the shimmering water going over the glass.

So, yes, our architecture is performative ... It performs, and it also works in an architectural way, an artistic way, in a sensuous way. Good architecture actually has to address both.

KOLAREVIC: It has to perform in both ways.

WHALLEY: Yes.

RAMAN: There are some fairly simple examples where the connection is quite easy to see. The design of a suspension bridge aims for something very beautiful and elegant but, in fact, every part of it is optimized for performance. The design of a concert hall is similar in its aims; for example, take the Kimmel Center for Performing Arts in Philadelphia and the interactions that went on between the architect who wanted to achieve a certain type of space and the acoustician who wanted to achieve a certain acoustic for the performance, and the theater consultants and mechanical engineers and so on. There was a process there that is simultaneously giving form and dealing with performance issues because ultimately the performance of that space is very, very critical. There are many other places where the definition of performance is a little vague and that is when sometimes we would go astray.

RAHIM: If you think about performance and performativity in terms of architectural space, you begin to think about cultural efficacy, organizational efficiency or maybe technological effectiveness. This is a really important issue and I think we have all been grappling with it in a disparate manner.

If you think about performance, it really depends on who you speak to with regards to how it has been perceived and how it can be influential on form-making. If you begin to understand each condition on its own and, in fact, if you ask any technical person how they define performance, they would define it in the way they do things; for example, it is 80% cooling efficiency if you are talking to a mechanical engineer, and the engine is 37% fuel efficient, etc. Then they follow it up with a very general definition of performance that is not quite that clear. Each of their definitions is localized to them.

Mahadev Raman, for example, places performance into a context specifically related to organizational efficiency. By moving away from the machine model into a network strategy, conditions such as quantities become qualities.

If we read performance with regards to form-making, it is really about intensity. As William Braham has mentioned, it is about an intensity of unseen vectors that are in the milieu, that then formulate and, at particular moments in time, become intense and produce an effect. It is really about intensifying the conditions so that there is a potential of producing an effect. If you zoom out and if you see the whole world with regards to performance, it is really about a cloud of contested vectors with particular intensities emerging. That is the only way to cut across the different paradigms of performance, shifting from the technical to the technological which is cultural.

KOLAREVIC: Ali Rahim talked about the notion of dynamic multiplicity. So if I could reduce, using the word "reduce" here operatively, the notions of performance to its essence, it is really about this dynamic multiplicities that are present, is it not?

LEATHERBARROW: That is related to what Andrew Whalley said about levels and the simultaneity of levels. I want to add perhaps a slightly separate issue. It seems part of what we have been struggling with is what yardsticks we have, what measures we bring to bear on the several kinds or levels of performance one expects of the building, leaving aside for a minute the problem of form-making or of design. Do we have a yardstick for beauty? Certainly not. Do we have one for health — a thermometer?

The concept of quality seems to plunge us into great uncertainty about measure. Part of the issue here is the spectrum of measuring tools we are willing to tolerate, from the most discerning and objective to those that are rather more personal and subjective. At some point the architect's authority shades off into that of the client, the constituency, the public at large. This happens when one says this room performs more or less the way we expect. In reaching such a conclusion, each of us invokes our understanding of common cultural expectations.

To quote Ortega y Gasset, the good architect is the one who knows more than an architect knows; the architect remembers what it is like to be in a lecture room, and the standard, the measure, the criterion of judgment, is not a matter of professional expertise but rather cultural knowledge. If we want to marginalize all of that, and class as performative those phenomena of architecture that can be measured according to the rulers of our expertise, I think we are reducing architecture.

We have to develop a concept of performativity that is sufficiently nuanced and subtle to embrace different kinds of measure. But I would not absorb all judgments of quality, intensity or beauty into what can be measured objectively, nor would I say that it is a separate issue from instrumentality. We should have concepts of instrumentality and performativity that sustain these different kinds of reading or aspects of measurement. That is where a number of the issues around prediction, accuracy, knowledge and foresight become rather difficult.

KLOFT: Other industries would run into problems if they could not measure formal, aesthetical and other aspects of what they create as consumer products. That is the difference between creating the product and what we are doing — the unique solution. Other industries establish mainstream rules as to what works right, so they do marketing, they create a world for the product — which we do not do or cannot do for a building.

Performance is not the same as efficiency; performance is much more than optimization. What we did in our freeform projects was not about having an optimized process.

KOLAREVIC: In my view, we cannot separate aesthetics from performance in architecture; the two are inseparable in many ways. We should not reduce the issues of performance to the things that can be quantified; it is much more interesting to discuss the intangibilities that begin to qualify performance in socio-economic and cultural terms.

We should perhaps frame the discourse of performance in a temporal fashion, i.e. what presently defines the performative in architecture; in other words, what is accepted as a structure, as architecture that performs well today, not only in a technical sense, in the sense of building physics, but also in a socio-economic and cultural sense.

What is performative today may not be so ten years from now, which goes back to what David Leatherbarrow has defined as *eventmental* architecture. When we talk about the events, however, we cannot really divorce thinking of the "unscripted" events from what is scripted in architecture — the program that the architects are typically given — because the program assumes or attempts to predict certain events. How do our practicing colleagues deal with these contradictions of scripted versus unscripted?

LEATHERBARROW: Ali Rahim said that part of the task of design is to recognize first what is inevitable and, second, the value of things over which one does not have control. The challenge is to see both as part of, not added to the process — both equally internal.

Once I had the pleasure of hearing Yehudi Menuhin perform in Teatro Olympico. It was a perfect instrument, it performed perfectly. But the genius of the performance was Menuhin's ability to modify it to the room. Speaking conceptually or epistemologically, one faculty the architect must possess is foresight — this technology will lead to that result. My counter position or complimentary position is that the architect must also possess ingenuity. In this particular circumstance, the performance has to be adjusted; you play softer or more loudly, more rapidly or more slowly. In other words, you have the technical procedure, the optimal solution, but the architectural insight says in this circumstance it can only be done this way, and that is part of the project; it is not added to it, it is not secondary to the initial so-called objective, repeatable and instrumental process.

Within the instrumentality of architectural work there must be adjustment and internal correction. The architect's acuity and judgment must be brought to bear on the particularity of the case. Ingenuity is always contingent, never planned, unpredictable, and calls for a certain spontaneity of the design. That is where I think the building or the design becomes performative in a much larger sense.

The concept of script or plan, which is what we call it in design, is inadequate to this second level of understanding, which is really when the project takes off, when it is more than what could have been done in any circumstance, because it is only possible and uniquely relevant in this one. When this happens the performance comes alive, the building starts to live and breathe with the client, with the site, with the time and the context, in a way that one is really excited about, as opposed to confident in. So one could put confidence at risk for the sake of relevance, immediacy, concreteness and engagement, and that is what I mean by unscripted, the particularity of the case.

BRAHAM: We are essentially using different languages to describe the same thing, which is the ability to say with some certainty what will happen within the dimension of, for example, producing a temperature in a building. And yet, there are much larger conditions in which, from every conceivable different position, that does or does not mean something, is or is not good; and there is confusion between that and much more narrow activity.

The difference emerges when we talk about structural or constructional performativity, which has to do with the durability of the artifact, and the other dimension of environmental performativity, which has completely to do with people's experience (maybe even not within the artifact). That difference manifests itself in teaching, as two different trainings in backgrounds.

It is a much more profound difference than just an institutionalized one. It is quite a different thing to talk about the object and its experience. It is far, far more dangerous when we start thinking that by adjusting the experience or the conditions of experience that we really have some effect over the people that come into a building. It is perhaps our hubris that we want to make people happy somehow when they come into our buildings.

WHALLEY: But that probably minimizes the problem to a certain degree. We can say we will design for people; perhaps one way of measuring the performance of a building is to try to understand the subjective interaction between the building and all people who are interacting with it. That might lead us to different things.

BRAHAM: I was thinking about Mahadev Raman's remark that commissioning actually made more of a difference in some cases than design.

WHALLEY: But the missing piece in all of that is the assumption that we can actually quantify precisely the experience of, for example, thermal comfort. Many papers have been written on this and there are design guidelines that have come to be accepted, but I know that when I am feeling a little ill the condition that I would find thermally comfortable is quite different to when I am feeling full of life and energy. So even when there are conditions that can be measured, the subjective response or the effect on the receiver of those conditions is not measurable in any personal aspect.

BRAHAM: The other connection I would posit is that performativity occurs with some kind of feedback between the designer and the milieu or the object and the subject. Although these notions of feedback are a much overworked concept, if understood in a much broader sense, it actually helps us somewhat out of the dilemma of trying to overextend our reach and understand the degree to which these things are constantly being shaped and made, both by the unpredictable events and by the ones that we have imagined will be accommodated.

KOLAREVIC: I want to ask our speakers to be specific about what has challenged their notion of performative architecture — in other words, was there something that you heard that you disagreed with?

MALKAWI: Let me clarify that a little bit. When we started, Branko Kolarevic had a certain idea about performance, and I did too. We could have had probably two separate conferences, but we thought the most important and interesting thing would be to explore the integration between the two. One of our goals is to try to understand the apparent "disconnect" between these two worlds, the scientific and the artistic, and the influence that computation has on bringing them closer together.

BRAHAM: What did you disagree on?

KOLAREVIC: The one thing that I find annoying is the engineers' love for the things that can be measured and quantified, and the designers' aversion to the things that can be quantified and measured. Some of our speakers alluded to this tension that exists between things that can be quantified and qualified, and things that are simply intangible. To me, this tension between these two poles is a very interesting territory to address and explore, and I am curious to hear what your positions are.

KLOFT: For some engineers, it is not so interesting to work on performative architecture because it is always optimized; it is much more interesting to work on a performative process and do some really extreme things, like we did, and bring techniques forward. So it is not only architecture that should be performative but the process should be too.

WHALLEY: I was really impressed by Harald Kloft's attitude in working with Bernhard Franken; you are given a form by Bernhard Franken and he says you cannot change it, you cannot touch it. It goes contrary to what you have been trained to do as an engineer — you are trained to optimize things, but he says: "I don't really need an optimized structure, I just want the form and make sure that it works that way." I would die if I were an engineer under these conditions ...

KLOFT: No, no ... you develop a new technique, a different focus. For example, on the "Bubble" project we concentrated on the realization, how to form the acrylic form, etc. It was very good for us as engineers to have a fixed geometry from the beginning because if it changed every week then we could not focus on how to deal with new techniques of realization.

KOLAREVIC: Fixing things in the process was necessary, was it not?

LEATHERBARROW: We have to put an end to this business of qualitative and quantitative; two worlds, black and white, good and bad — forget it, it is not true.

KOLAREVIC: I never said that either one is good or bad.

LEATHERBARROW: They are not two, they are one — it is one spectrum of decisions, more or less certain. How do we judge whether or not this lecture room is too bright or too dark, too warm or too cool, its construction a waste of materials or not?

We make these judgments because we have been in lectures before; we have been in rooms like this, and each of us carries with us a whole history of cultural background, of experiencing lectures. There could be a measure or a spectrum of measures, some very precise, some reasonably so, some barely at all, and design works with all those levels. Nobody prefers quantitative or qualitative; design intelligence knows which kinds of things require which kinds of measure. The profound mistake is to try to apply this kind of measure to that kind of thing, as if there could be a yardstick of beauty. But that does not mean to say we could not look at this room and say it is pretty ugly or it is marvelously beautiful; one could say that there would never be any agreement about that, but I would say there would be more or less disagreement. If you could tolerate degrees of certainty, and if you could see design wisdom as knowing which kind of problem requires which kind of certainty, then I think you can avoid this polarity, which I believe we should do. If we want to talk about the performance of a space, we should see it as embracing that full spectrum of measures. It is just that some are quantified in one way and others quantified in a different way.

So if it is true that there is something like a cultural norm that is called the lecture theater — and we all basically know what it is (you sit there and we stand here, that is darker, this is lighter) — we can say this room is more or less correct with respect to that norm. Is it fixed? No, it has to change in each case, and those changes give design its freedom. My view here is that you are completely wrong when you polarize qualitative and quantitative.

KOLAREVIC: I perfectly agree with what you say. I am not arguing for either of the poles, and actually I am not separating the qualitative from the quantitative. I am pondering the worlds of engineering and the worlds of architecture, and how they are commonly perceived and what they stand for. Architects love to think that they deal with the intangibles and the engineers love to think that they are dealing with the things that can be quantified and therefore qualified. Those are the polarities that I am trying to establish.

WHALLEY: Engineers like to work in vectors, do they not? A scientist likes to do an experiment and if the experiment produces the same result ninety times then that is a success, whereas artists and architects work in circles; we can design something a hundred times that appears to be the same, yet to the artist or the designer it is a different exploration each time. These are the two poles that we are trying to blend in architecture.

LEATHERBARROW: You have the scientist and perhaps the artist at the two poles, you have the architect and the engineer moving substantially in from those poles into a middle ground where you cannot be as extreme as you implied in your statement.

KOLAREVIC: The design practices that manage to somehow operate in this middle ground between the two extremes are the ones that are successful in the contemporary context. Grimshaw is a practice that occupies that middle ground between the polarities I have described. Harald Kloft's engineering practice is willing to accept this fuzzy ground that architects have now offered them as a given — without questioning?

KLOFT: We are not interested in measuring things. Our aim is to support the architecture, to provide a performative process for a performative architecture.

Equally important is the issue of adaptability — buildings cannot change like other objects. In the future, we should look more into this issue of change, i.e. how we can deal with the vectoring, the climate concepts.

LEATHERBARROW: I disagree, buildings do change.

WHALLEY: Architecture that was traditionally quite static is becoming more and more active, adapting itself to the seasons and to change. You can have an environment where, when the summer comes, you can allow the air temperature to go up a bit, but you brought in the shading and the sunlight permeates through that shutter. That speaks to the qualitative — it is a sensuous thing. You are in touch with the season. The building is performing in a certain way.

I suppose the antithesis to this kind of high performance — performative — architecture would be a box where you switch on the air conditioning and set it at 20°C and it holds it winter to summer, which is what has happened in the 1950s and 1960s. Whereas now, architecture is performing in interesting ways. It uses architectural devices in the same way a boat can lift up the sails and capture the energy of the wind. Architecture now responds in similar ways, as one can see it in traditional vernacular buildings. It always has had that location-based response to where it was placed. That idea of placement in architecture has to do with the place and the season, and so on. That is when architecture really becomes a rich experience.

RAMAN: The approach previously used was a kind of objective performativity, where you are trying to achieve certain measurable goals, or certain styles, or certain results. The new approach to designing environments is much more of a subjective performativity. So, if you say that the occupants ought to be able to adjust the shading, the light level, the temperature, etc., to suit whatever their particular mood or requirements happen to be at the time, then in a sense that becomes a little more precise as a measure of performativity that you get if you achieve that flexibility in the space.

LEATHERBARROW: Absolutely. That kind of attunement is more precise than the optimization which is indifferent to the particularity of the circumstance.

According to Cicero, a good public speaker is the one who can adjust the speech to what the audience is capable of understanding. This judgment of what is right in these circumstances is not subjective — it is situational performativity. In particular situations, performances unfold in particular ways. When the context varies, the performance cannot remain the same. It will be ineffective. So the challenge is to optimize in a given context. It is not optimal in itself as such in any possible application; rather, the real architectural optimization is a situated performativity.

Architecture must engage the contingencies as part of its rationality — a contingent rationality, a weak rationality, a situated performativity. That is actually more precise than the precision of the suspended suspension bridge. It is nowhere; it is really brilliant at nothing. You might call it a meaningless certainty.

MALKAWI: There is also the question of computation. Does it help the process or not?

KOLAREVIC: David Leatherbarrow says it is a distraction.

LEATHERBARROW: No, just the reverse. But do not try to do everything with just one instrument. Do not try to look through one pair of glasses to see the whole world. The computer is brilliant at what it does, but to make a beautiful building out of a computer, forget it. Just forget it. There are many other things you need, and it will not help you with those things. It is like Le Corbusier's Modulor — it helps you with some things, but for the really difficult things it is useless.

MALKAWI: Let us try to think about the development of computation within the past twenty years. In addition, let us take a look at buildings in relation to their performance. Is there a link between the development in computation and the current "high performance" buildings that we see today?

RAMAN: I see that the intrinsic process has not changed. What computation has done is that it has had a liberating influence on design. It is not that the computation takes over design or replaces it in some kind of way. What it does is it allows you to explore more — more features, more materials, more situations with a greater degree of confidence than one might not have been able to do without. But it has not intrinsically changed the process.

WHALLEY: One area the computer has impacted on is the idea of three-dimensional form in design. It would have been very difficult twenty years ago to do some of the structures and forms and shapes that we do now. You could have done them, but probably not as efficiently. The computer has slightly expanded the repertoire, or allowed certain areas to be explored, which before were just not open. So it is a tool. It has broadened what we can do, but at the end of the day it is a tool.

KOLAREVIC: Let us then address the liberating dimension of the technology that several speakers have referred to. As Bill Braham mentioned, the air conditioning in the 1950s was thought of as being a liberating technology. If we accept that what the digital now offers us is another liberating technology, I wonder if we will be regretting these new liberties at some point down the road; maybe not. I think we need to ponder these questions as we embrace the technological advances with enthusiasm.

ROBERT AISH (from the audience): I take up the point that Andrew Whalley made about the impact of computers in this process, in particular in the geometric quality. The next stage, which I think might be more interesting to discuss, is that, essentially, in creating these computer tools, we are looking to generate a representation that can be shared by both the creative architects who are dealing with subjective aspects of the performance and the engineers who supposedly deal with more measurable aspects. If we can create tools that allow communication between the creative side and the engineering side, I think that will hopefully have a very positive impact on this whole area.

MALKAWI: My argument is that computation within the past twenty years has tried to bridge the gap between the artistic and the scientific. Computation is slowly integrating those two or at least trying to bring them together by enhancing communication between those two extremes.

AISH: One really needs to go further than that. As Andrew Whalley mentioned, the scientist is doing an experiment over and over again, in comparison to the artist, architect or designer, who is inventing every time. The role of science or engineering is not just to repeat the experiment, but surely to produce some causal model, some explanation. If we can build this kind of building engineering causal model into our design tools, then these can be directly accessed and inform the design process. We can actually build the prediction of the experience into the design. That is what completes the loop because essentially we want to design with feedback.

JEAN-FRANCOIS BLASSEL (from the audience): Ali Malkawi has asked if computation had brought about some changes. There were no computers in architecture schools twenty-five years ago, and now there is only computers. That is not just a leap in the quantity; we have a leap in the type of computation that one can do about the physical behavior of a building. What one can do now is, in fact, to build a simulacrum, a simulation of a building, and experiment with at least some aspects of it. Whereas previously, the main thing one could do is analyze it and then size it, give thicknesses, give types of materials, and so on. Today that is quite different. You can take a bunch of sticks or a strange shape and you can try and decide what is going to behave as a structure, which is a very, very different position because of simulations of a different sample. This has really changed the way one could look at technical issues, and at the same time it opened up new possibilities for architecture and other fields, as we could now build strange shapes. We also see the changes in the way one can explore the thermal behavior of unusual spaces.

All the fields were very good at basically perceiving only one solution. You could only use a solution that is already known and adopt it. But now we can play with something and see if it will work. There is a "black box" danger in this, but clearly something very new is happening. We are trying to find out what to do with the new computational technologies. That is the crux, our real goal.

CRAIG SCHWITTER (from the audience): While everybody has a computer today and there is a tremendous computing reservoir at our fingertips, the models are still incredibly crude — we are in version 1.0. We can model extremely complex kinds of structure surfaces, have all kinds of cutting patterns, and yet what we have actually seen is a very simple, elastic relationship with material properties. Things are still very crude.

Somebody mentioned that the juxtaposition of the quality and quantity is interesting. It is quite stereotypical of architects and engineers who use it.

Some of these more complicated geometries are sometimes actually driving deeper wedges between architecture and engineering. We often do not have the ability of doing a feedback loop on a project. Sometimes architects do not even want to have feedback. Sometimes these performative techniques actually lead to extreme divisiveness. It is nice to say that we all work with the same model, but those of us who worked on multidisciplinary projects know that we all parse off into doing our own little models. It is rare that fabricators actually use the model that we give them. They might use it as a basis for their explorations, for developing their own models that they can trust. So, sometimes these digital technologies can actually be quite divisive. As we look for things that can draw us together as architects and engineers, we need to be very conscious of that.

MALKAWI: It can be dangerous to a certain degree if one uses the computational tools as a black box without really understanding the way it works.

SCHWITTER: I do not think one uses the finite element analysis when one is designing a structure. I do not necessarily prescribe to the black box theory. Sometimes I think it is actually acceptable to use computation as a black box as long as one understands the greater discipline that one is working within, in terms of being able to ask whether it is the right solution or the wrong solution.

RAMAN: I would qualify that slightly by saying that the black box sometimes takes the place for an innate understanding of what is actually going on. An excessive reliance on the computer sometimes detracts from developing an intrinsic field (or perhaps something worse). What you are talking about is that you are not doing a finite element analysis because through your experience you have a feel as to how the structure works and what is likely to work and what is not. I have sensed that for the new generation of engineers coming through that understanding is being replaced by the reliance on this process.

LEATHERBARROW: My only point about the computer would be not to mistake the "menu" for the "meal." I do not think the representation is the same as the reality. That is the only issue; that is partly what Ali Malkawi is struggling with — the degree to which the feedback from outside the system is adequately represented within the system. Now we can build up and make increasingly more sophisticated representations and simulations, but at what point do we just forget the meal and eat the menu? The representation, the simulation, are absolutely necessary, powerful, effective, unprecedented — all of that is true. But it is still partial and there has got to be somehow, within the system, a recognition of its relative autonomy.

The question of engagement seems to me the most pressing one for performative architecture. The simulation is adequate under certain circumstances but when those circumstances change, the whole framework needs to adjust or modify itself. It is the question of limits — how much of this so-called feedback can be absorbed into the instrumentality.

I do not think there should be any artificial limits imposed on the research. In fact, there should be more and more simulations. But I think there should also be a sense that what has not been simulated will actually enlighten the project in the performative sense one is after. But maybe we differ on this.

RAHIM: With regards to simulation and developing and evolving projects, it is an emergence between the environment and the user. That is in a certain sense similar to knowing what you are working with. For example, if you are working with a hammer, you become a hammer, and if you are working with a temporal tool, you have to become one with it — and that is exactly what non-linear software allows us to do. It is to provide and provoke many conditions that you could not foresee without actually engaging the development of the process.

LEATHERBARROW: What about the representations of the conditions you could not foresee? Mine is only a question about the adequacy of the representation to the given condition.

RAHIM: That is the question. If you think of the difference between the virtual and the real, that is where there is some very interesting potential because you have to go through the process of actualizing something with virtuality. The process of actualization is a mediation process with which we operate in a given moment in time.

The representation here itself is not limited because one can run a hundred scenarios and the formation produced contains these virtualities which are multiple and not singular. The goal is not to reduce these virtualities into the singular, but to maintain a multiplicity in the formation.

ANDRE CHASZAR (from the audience): There is a relationship between simulation and exploration. So when one talks about performative designing, often it seems that the desire is to arrive at some sort of emergence. That something would come out of it that one had not foreseen. Is that valid from the point of view of engineering analysis? If I understand Mahadev Raman's point of view, you should know the answer before you begin, and so when you run your simulation and you get a result that you did not foresee, then your first thought would be that the simulation is done incorrectly.

RAMAN: I had exactly that experience within the first two years of practice where I was the new kid who knew about computers and punch cards and those things. So having gone through an elaborate simulation for one of the Richard Rogers' buildings, I proudly presented my cooling load calculation freshly off the computer to a crusty old engineer who said I was out by a factor of four. Sure enough, I had made some mistakes and we eventually got to the point where I got it right, both to his satisfaction and to mine because I did actually discover the mistakes. But that it is not entirely clear-cut because sometimes the solution emerges through the exploration of different scenarios and the optimization process. There is also the ability to recognize that solution as it emerges, which is something that is outside of the simulation and is innate in the experience of the practitioner. The way in which a problem is presented to you in engineering school is: if this, this, this and this, what is the answer? You use a little bit of experience and do the calculation, you come out with the answer and the professor can tell you whether it is right or wrong. When you go to the design office, working with a group of engineers and architects, you find that what you do not know outnumbers what you can define. So in theory you are right. If you could define all of the parameters that go into a process, and if you have a simulation of the universe that is correctly dealt with, you could set that off and eventually come up with something that says this is the answer. In reality you can do that, but you can do that only within very limited areas where the boundary conditions are definable within limits. Frankly, I do not think we will get to a time, in my generation, where you can model all of the complex issues that go into design in such a way that the computer can come up with the optimal answer.

KLOFT: You need a lot of experience dealing with these programs and these tools. On the other hand, architects and engineers are coming closer together with these tools, as we are experiencing in our office, where we have both architects and engineers.

Engineers should not only be educated that this is right and this is wrong. We normally cut sections through the building and we do not think like the architects in a spatial system, which is especially true for "freeforms." In our office, much of the work with freeforms is done by architects. So, there is a chance that the tools can bring both closer together in the future. But I do not think that one tool can bring the whole solution.

BLASSEL: Some of the models can cast more light on some aspects at the beginning of the design. Perhaps architecture is not really so much about the space as it is about judgment, about the synthesis. It is about taking a whole range of questions, weighing different things. Where these computations become interesting and so confusing is the fact that you have to bring in all these things and decide which is more important — the energy efficiency in the building or views from it, etc. There are just so many things to consider. The computer models may help us decide. They may help us to understand better.

18

OPERATIVE PERFORMATIVITY

FRIED AUGENBROE
JEAN-FRANCOIS BLASSEL
JAN EDLER
PETER McCLEARY
GREG OTTO
LARS SPUYBROEK
ALI M. MALKAWI
BRANKO KOLAREVIC

18

OPERATIVE PERFORMATIVITY

FRIED AUGENBROE
JEAN-FRANCOIS BLASSEL
JAN EDLER
PETER McCLEARY
GREG OTTO
LARS SPUYBROEK
ALI M. MALKAWI
BRANKO KOLAREVIC

MALKAWI: I want to reflect on the original idea of how we came to think about this topic and to briefly introduce the subject that we will discuss in this panel.

Performance in architecture is not new and is a very general concept that has been discussed in different forms. From a theoretical perspective, the concept of performance has been discussed in this symposium from phenomenological and structural perspectives. Our invited speakers come from a variety of disciplines and we tried, as much as possible, to select individuals who would bring various views from the conceptual and operative sides of performance. In his presentation, David Leatherbarrow provided a comprehensive introductory framework for performance from a theoretical perspective. The duality of the subject was enforced by Peter McCleary when he discussed issues of negotiation between the opaque and the transparent, and between art and science. He provided a theoretical framework for reflection on the complex nature of architecture. We have witnessed this negotiation or mediation between components being discussed in a variety of forms during the symposium. One of the main issues discussed as a possible factor in bringing new forms of architecture is the computational instrument which I believe can act as a mediator or an interface between these two worlds, art and science.

Fried Augenbroe talked about the issue of measuring performance. This is a topic that we have been working on for a long time. The topic has a long history from theoretical as well as practical perspectives. One of its byproducts is the computational instrument that has been used in design and analysis. Although there has been major advancement in this area and we see its influence on the profession, there has been a "disconnect" between commercial tool developers and the designers.

We see two types of users, the architects as well as the engineers or the consultants. We do have tools to answer specific questions from an analysis perspective, but we lack the tools that aid in synthesis. Our tools also lack the understanding of the process that supports the design analysis as buildings are designed and constructed.

We discussed performance from an aesthetic and communicative point of view, such as the façade project. Discussions also included the cultural importance of performance in regard to buildings. The Buro Happold group talked about tools and their adequacy and validation. Do performance numbers provide an absolute measure which is always correct? Different problems require different solutions and, in most cases, measuring performance can be comparative; relative performance is what is needed in many instances.

Lars Spuybroek described the measurements of motion in relation to space; Jean-François Blassel addressed the coupling of different types of software to answer questions of performance. Finally, Peter McCleary provided a perspective on how the different actors of the design of a building interact although they have different languages.

We attempted to separate the discussion in the panels from the operative as well as the conceptual perspectives, although we understand their interactions. The operative component, which is the topic of this discussion, is related primarily to building operation and its tangible measure of performance, a performance that can be judged by relative benchmarks. Operative performance, as we see it, is closely linked to the computation instrument and this is what I would like our panel to focus on discussing.

The first question for the panel is related to the instruments or the analysis tools and how they influence design. What is the role of the analysis tools, as it stands now, in influencing design performance? Try to reflect on it from your own domain of expertise.

AUGENBROE: I have no direct design experience myself, but I can reflect a little bit on your question. You and I, and many others, have been working for many years on improving building simulation tools and we have never reached a more mature stage than we have at the moment. At the last international building simulation conference, where the latest tools including CFD (computational fluid dynamics) were presented, you could conclude there was very little in terms of functionality that was still to be desired. And I remember, in the closing session, one of the overriding conclusions that we came away from that conference with was that simulation is so mature it can eventually become invisible.

So, that leads to the next point and that is how does simulation then support design? I think, at the moment, the only answer is: in a very limited fashion. And why is that? I think because of the lack in formality. We use simulation tools to study behavior during design evolution, but we do that in a way that is not well coordinated. My conclusion, as one of the people who wants to look beyond what we do with current matured simulation, is that we need to develop something which puts simulation and analysis in a much more formal framework. What I suggested in my presentation was to base this framework on predefined virtual experiments. We need to define clear expectations of the design and be able to measure proposed solutions against those. For this we would need very matured simulation tools, which have to be preconfigured in a way that you can perform these virtual experiments. I have not seen any platforms being developed to do that in an effective way, but I know many efforts in research that are trying to do just that, leading to what you could call the next generation of simulation.

BLASSEL: The tools that we have now are about making abstract presentations of some aspects of the building. But this is not a new situation. Currently we have specialized tools that are part of a toolbox. We do apply them to specific problems. As Fried Augenbroe suggested, if there are similarities between projects, maybe we can call for an array of tools that can be available and conform to a benchmark which allow us to compare solutions.

But for the generative part of the project, I would think that we have to use tools in a very discerning fashion to develop knives with which we cut the problem, just as all architectural problems are very complex, with all the issues that are interrelated. These tools allow us to make a very minute dissection of what we are trying to understand. Depending on the shape of the problem, we need to cut in different places using different tools, so we need to find specific tools to help us imagine solutions.

KOLAREVIC: I would argue that the current simulation tools are completely useless from a design perspective. If you do agree with that as a position, then you ask what will make them useful. What could make them useful is perhaps changing the resolution. High-resolution tools demand an incredible degree of detail, if you want to get a useful simulation from an engineering point of view. We need some kind of low-resolution simulation tools that are approachable by architects.

The conceptual design is not about detailed modeling of the rooms so that you can do the CFD simulation. If we need low resolution, then we would somehow need an interface that does not demand such a high degree of domain knowledge. You almost have to be an expert in environmental design before you could use any of the CFD software. What we may want to see from a design perspective is something that has a genuine dimension from the performance point of view. So, coming from a design perspective, what will make these tools useful to us? Right now I do not find the analysis software of much use in conceptual design.

BLASSEL: Well, maybe what you are trying to say is that the low-resolution tools require more abstraction — it is like a sketch before we do a design. What we need is the equivalent of a sketch, a technical sketch.

MALKAWI: This might be related to the larger question of the user of the tool. Is the user an expert or non-expert? Is it the engineer or is it the architect? Much debate has been spent targeting this question during the past decade and the consensus is that the simplified tools did not deliver their promise of helping the designer. Is it more productive to understand the process of design, including the collaboration and communication between the different participants, and develop tools that support these activities better? Identify the users within the process and accordingly develop tools that satisfy the question of resolution that support these users but do not focus only on the architects.

SPUYBROEK: Every project starts with analysis and with the analytical and the weighing of data; basically, you do that with every project. At the end, the project also ends with things that are weighed and that actually cost money. But, in between, there is much non-metric stuff going on. I think the whole art is actually to move from the real that is existence, to another real that is the real plus your project. But you have to pass through something that is not real or that is actually abstract. Of course, that is the whole question of the diagram and the different shapes in between organization and structure.

So, we would have the world of organizations or diagrams and you would have the world of structures that actually can be quantified. But there is this passage, it has to pass through a stage of order and qualification on an abstract level. The question is how do you go from tools that are only capable of cutting things up to a tool that can actually synthesize and integrate and reach an abstract level.

MALKAWI: On the note of abstraction, I think it might be worthwhile to ask Peter McCleary about his thoughts.

McCLEARY: Aristotle said something similar on the subject of skill. He believed that one should become sufficiently informed to understand the playing of the musician, the rhetoric of the poet and the "how" of the builder. If one becomes too skilled, it is not possible to differentiate between the master and the slave, and it was never our intention to be a slave to the mediation.

Recently, after my short talk on the relationship between structure and space, one of my distinguished colleagues said that there was no longer such a relationship. In this era of Frank Gehry and other Baroque architects, the building's internal surface shapes, or is shaped by, the activity space, and the external surface derives its shape or configuration from the geometry and scale of the city context. The structure is sandwiched between and ranks lower in importance than the spatial relationships for the inside and the outside.

There is a proportional relationship among the three dimensions of the volume of the space framed by a structure. This does not mean that the height is the harmonic mean of the two dimensions of the plan, as it was for Palladio. In plan, we weave the warp and the weft of both the space and the structure; their proportional relationship can be described in the language of mathematics in general and geometry in particular. Just as we can speak of the geometry of the space and of the structure, we can describe the geometry of sound and the geometry of light. There is much work to be done on the synthesis of those geometries. In the absence of operational techniques, it will remain the task of mature designers to negotiate and coordinate that synthesis.

KOLAREVIC: That ties back to the issues of representation that were touched upon several times. The issue of representational integration relates to the issues of geometry that you are bringing up. That is another significant missing element in the visual tools that we have access to. Perhaps these different geometries that Peter McCleary is referring to could somehow be brought together in this visual domain. The reason I bring these up is that I was surprised by Fried Augenbroe's assertion that we are done with building simulation. His suggestion is what we have is already sufficient and mature. I think there is quite a bit to be done in that realm.

MALKAWI: Knowing the work that is being done in the field of building simulation, which is not just limited to developing tools for the architects, I would like to give Fried Augenbroe an opportunity to respond and talk about the representational aspects of the simulation and the work that he has done on interoperability.

AUGENBROE: I would like to respond to your earlier statement that tools are completely useless. No, I do not agree with what you said. I think you were close to saying that tools are completely useless to designers, but you said they are completely useless to design.

KOLAREVIC: That is correct.

AUGENBROE: Which one did you mean?

KOLAREVIC: Both.

AUGENBROE: I think today we have many tools that are certainly not useless to design. Whether they are useless to designers is a loaded question. There is an old discussion stating that we are placing the wrong tools in the wrong hands or we are making the wrong tools for the wrong purpose. I think we are actually still doing that.

One of the things that is haunting us in building simulation research is that we are developing functionally undefined tools. We are so focused on studying the physical behavior that we are first and foremost trying to capture that behavior in our tools, and then count on the expert and whatever divine inspiration occurs in a given project to use the tool effectively. Reality has it that tools are used regardless of whether a certain simulation actually makes sense. And that is where Branko Kolarevic's statement comes from, is it not? If so, I agree with him, in a sense, but at the same time I am amazed to hear designers requesting tools that work in conceptual stage and maybe later on too. This is in contrast with a movement I see going on in corporate America where more and more firms are delegating rather than incorporating the expert engineering analysis work. So, why would design firms try to go against that trend — even if the tool is very simple? I think that designers will stop asking for simulation tools once they realize that if you have a trusted medium under the next mouse click, this medium will give you more accurate information about the behavior that you are trying to study to make the next step in design, and more quickly than you could ever have generated yourself.

MALKAWI: I would like to give Greg Otto one last word before we move further.

OTTO: I would like to try to link together the interesting conversations we have had so far. I think design is actually changing and I think we have changed into a more collaborative state. I am finding that in many projects that I get involved with now, the architect has an idea, it is only an idea and now he wants to sit at the table and he wants to sort out the design. In reference to the CFD model or the simplified version, I do not know if that is exactly what you want. Do you want a machine to sit there or do you want somebody sitting at your table with a world of experience, informing you with what are the merits or what are the downsides of this, that or the other? So, with regards to tools, I think that the current tools that we have allow us to look at more scenarios and they are going to only get better. But, you will always need the people to interpret the results and that is the round table effect. I also believe that this is going to lead to a performative architecture because it is influenced by a world of knowledge, not a singularity of knowledge.

MALKAWI: At this point I would like to invite questions from the audience.

UNKNOWN (from the audience): I am not sure if we lost track of that idea of the usefulness of lower resolution analysis or simulation tools. I hope that we have not. I think Greg Otto correctly pointed out that you want to do this in a collaborative environment. So, if you are developing a lower resolution analysis tool, the purpose of doing so is not to enable the architect to do the engineer's work. But it would actually help the engineer fill that collaborative role earlier in the process. So, conceptual design in the absence of those tools still happens in the brain and on the paper. It is only late in the project, often too late, that you can leverage the power of these analytical tools. You can make them simplified to the point where they can be used without knowing too much already, but still getting useful answers. Then you can get a better use for the tools.

MALKAWI: We have two issues on hand: the tools that are related to the process and probably different types of resolution and the interoperability between the tools themselves that will support the process itself. This has an effect on the issue of accuracy required by the tools. I would like to call on the panel to get their thoughts in regard to these issues. I would like to start with Fried Augenbroe as he and I have been working on these issues in our research.

AUGENBROE: It is not for lack of trying that we do not have tools that do something meaningful in early conceptual design stages. A few years ago we did an interesting project to find out whether current tools are accurate enough if you look at them from an uncertainty perspective. We found that there is a technique called probabilistic inversion where you try to find out what is the simplest tool that would still give you the same accuracy in the results given the fact that we work in an uncertain world with uncertain input. We then proved that most of the tools could be much simpler than the ones we are working with. So, given the fact that your input is very uncertain, a simpler tool may give you the same level of resolution as the more accurate tools. This is an interesting and somewhat uncomfortable conclusion for tool developers. Now, how does that translate to making tools simpler or having a lower resolution (which, by the way, are two different subjects)?

The 1970s were dominated by attempts to develop simplified tools — dumbing them down, so to speak, so they could be used by people who were not experts. I think that movement has virtually died down. A more interesting area that we are working on now are tools that still reflect the basic physics of the field and are not necessarily that easy to use, but require much less detailed input. Now the problem is that if you look at the uncertainty levels in the very early design stages with regard to information availability, you will find that you hardly get any resolution from these tools to support design decisions. If this is ignored, they become very dangerous because they deliver some kind of quasi-certainty.

I think that we should do more work on defining simplified experiments that can be applied in the early phases of the design. They should require only very limited input, so you have to carefully design those experiments, and then make simulation tools to execute them. That, in my view, is the way to go, not taking the simulation tools that you already have and trying to make them so simple by defaulting and doing other kinds of things to them to make them usable in a very early stage when you have much less refined knowledge of the input data. I think that will not work. So for the first time we are in the stage where we can define the kind of performance quantifications that we are looking for. It is really back to the drawing board. But the point is that all the projects that we saw debated at this symposium are so far out of the routine that it will be an illusion to think that the development that I described will give you any useful support for those kinds of projects in the near future.

SPUYBROEK: We are discussing tools as if they are towards determinism or to a purpose. Marx, of course, already knew that a tool was not for the user but actually producing a user. That is the first product of a tool — its user. It is not that we were users and then all the software companies are designing software for us. It is not true. It is actually the other way around, and that is why I want to argue a bit against instrumentalism and argue against this whole idea of mediation where there is A, there is us, and then there is the world that is B. We just mediate and bridge phenomenologically with our tool that is neutrally hanging in-between. I do not think that is true, even for a hammer; something so overly defined by its function to hit a nail in the wall is every now and then used to kill somebody. So that is already going beyond the profile of the thing.

MALKAWI: I think we are not articulating our needs as designers to the industry or the tools' developers. This is another discussion; however, I would like to ask Peter McCleary to respond to Lars Spuybroek regarding the issue of mediation.

McCLEARY: In this context, there are at least two interpretations of the phenomenology of tools. While humans and their equipment are part of a contextual totality, tools are either equifinal, that is different tools can serve the same end or purpose, or are equipotential, that is the same tool can serve different purposes. Thus, a hammer does not insist that the world is a nail; or, as the Chinese say, it is the one that sticks up that gets hammered on the head.

KOLAREVIC: What are the kinds of potentials and opportunities, and what are the pitfalls that we should try to avoid when thinking about the performative in architecture?

SPUYBROEK: Architects never see themselves as specialists, and that is basically part of the problem because they are normally at the table as generalists with specialists helping them. The specialists are dealing with the hard data and the architect always jumps on the table and puts it all together, but nobody really understands how. That is difficult. It is not about having the engineers conceptualize and going up a level higher, so they can actually conceptualize with us during the early stages of the design. It is actually the other way around.

AUGENBROE: The last two questions are closely related. What we are really probing is the way you see the whole building process being more managed and what is the role of the architect in that, and what kind of collaboration environments will ultimately work as a catalyst to change that process. The interesting part for designers is how to anticipate a future where, indeed, collaboration is the driver of the process. You get into issues of formality, predictability and coordination, based on integrated representations, linked to integrated performance taxonomies. We have to prepare our profession for this future, if indeed that is the future, of very tight and coordinated integration with all of the other involved disciplines. If the architect wants to play the role of the design manager — as I think he or she should and might want to — the architect has to come to the table with the tools and representations that allow him or her to manage the performance contracts, making explicit in what way the engineers have to supply the right expertise. I think these two questions point to a direction where I think performative architecture should go; keep a close watch on what is happening in the industry in terms of collaboration contracts and project delivery systems.

UNKNOWN (from the audience): I would like to know whether we are asking too much of the architect or asking too little in how much we are requiring him or her to do with regard to deciding how a project moves forward. For hundreds of years, perhaps thousands of years, we were able to design buildings that met their performance objectives. We were able to do things reasonably well, and perhaps we were able to do that because of the types of materials we used. We did not necessarily design things that would not work. It seems now that we are able to design things that would not work and we have to hand it to an engineer who is going to undo the problems that we have created. For example, it seems to be ridiculous at times that we design glass boxes and put them in the desert and ask our engineers to find a way to make them work. In some ways are we doing such things because we can do things that we could never do before? Are architects now only generalists and so are not contributing much to the profession?

McCLEARY: I have been lucky to work with some architects who were both generalists and specialists. At the meeting of design collaborators, the architect had the broadest and deepest culture. Not only were they generalists but they also had particular expertise. At that meeting it did not make sense to focus on or even talk of the percentage of lime, silica and alumina in the cement. Rarely did we reach that level of detail. When we did, no other person understood its significance. All participants were asked to bring something that could be synthesized into a whole. Since others would not understand, the architect might not talk about mood and feeling. However, they could define the characteristics of the activity and what kinds of spaces might accommodate that activity; even a space that "inspires an activity." While he or she might not know the measure of the space, he or she had a good sense of the orientations or isotropy of the space. In addition to space planning for activities, the architect might propose a new structural configuration, or even a new way of living in space. Thus, the architect brought a considerable degree of oversight expertise to the collaboration. Their proposal was not a building, represented graphically, but concerned those things that have to do with habitation. No one else did that. So, yes, the architects were generalists but they also brought a lot of particular skill and knowledge to the table.

AUGENBROE: Getting here, I noticed that the main focus of the designers was to bring their design to the public stage and have it perform. To me, that was significant. This showed a kind of bias in the choreography, with focus on seeing the object perform on the public stage rather than regard performativity as the capstone for the matching of client expectations and fulfillments. From my perspective, I think that managing design choreography is ultimately more important from a research and dialogue point of view than having the building as an object perform on a public stage. But that may be strongly biased by my engineering background.

BLASSEL: I would like to go back to the question of the glass box in the desert, but not so much to provide an answer about the glass box but about what is going to follow. That is, why can't we just do it the way it has been done before? Maybe it is because we find ourselves in a global position and, at the same time, find ourselves in a comfortable situation which we are now enjoying in the United States or in Western Europe. But we do not have enough resources; something needs to be done in a much more inventive way using the resources more cleverly, using energy more cleverly. I am not saying that technology will be the only way to address this problem, but it is one of the ways in which it can address the problem. So, the tools, and the fact that we have to try to use them and try to make them evolve and understand the way in which the physical nature of the building interacts with its more spiritual nature, is essential.

EDLER: One of the main issues Peter McCleary discussed, and that I agree with, is communication. This is an aspect that we did not discuss much and it is blocking many powerful materialities of architecture right now from appearing. It is a very pragmatic issue. As I started as an inexperienced professional — a typical architect coming from university at some point hitting my first real construction site — what I experienced was that energy gets lost in architecture at the point when you try to erect it. Suddenly, you have a situation which is purely about competition and being responsible for something. Who is the person to take responsibility? This is a very crucial issue because, as I remember on the construction site in Graz, the main topic was basically trying to take as little responsibility as you can because it is all about money and this is causing so much friction. If there would be a possibility of getting this friction away from the production of architecture, I think architecture would be performing much better.

McCLEARY: Even though everyday the performance aspect of each technique becomes more resolved, I am not persuaded that everything is measurable. I am not persuaded that things that have not been measured are capable of being measured. One of the architect's responsibilities is to reveal aspects of our environment that seem to be hidden or concealed. For example, water can be fixed in place or I can carry it around with me — to drink or it can be conceived as H_2O. It also has a meniscus, it has depth, it has pressure, it can flow, and so forth. There are many ways to interpret an object. One of the major responsibilities of the architect is to reveal, through designing and building, the nature of humans and their worlds and the dialectical relationship between them.

SPUYBROEK: Today, performance is related to form and information. I see performance as something that comes after function and after the event. We had years of function where bodies were mechanical and mechanistic. You had necessity, you had necessary behavior, mechanistic behavior, and then you had play. You had this space of accidents where you had play and multifunction. I think performance is a real issue in the sense that it is trying to look for actually merging these. Because we lost some time during postmodernism and deconstruction where architecture became a language, now there is a real interest. All the conferences are about electronics or about the body, and there is a real sense of materialism. It is not a reductionist materialism but it is looking at the real materiality of experience, feelings, being and structure. What does architecture actually do? What is the operationality of it, instead of its aesthetics and its language? I think this is performance but I understand there is something in between which is mono and multi, something that is not either deterministic or totally wild but is actually some vague, wider sense of redundancy. Not free play or pure function, but is actually a widening up of function that makes it relate back to bodies and to structure.

OTTO: Peter McCleary's remarks provide for a very good closure. The world that we face is growing more complex and I think in that complexity we are forced to deal with more issues and, as Peter noted, we can measure the scientific but there are many other things that are not scientific and cannot be measured. To me, performative architecture is trying to find ways and methods to understand and deal with those issues in a way that, ultimately, ends up with a result that is satisfactory to the design intent.

MALKAWI: In my opening remarks for the symposium, I stated that we tried to organize it not only with the hope of raising questions regarding performance, but with the attempt to find common threads between the different views presented regarding performance. I think we definitely raised the questions and we found common threads at least conceptually. We did not and we are not trying to synthesize all the different aspects and views of performance as being projected from our discussions. As many of you stated, both David Leatherbarrow's and Peter McCleary's presentations provided the needed anchors to introduce and close our event in regard to performance in relation to architecture. What I would like to do now is give my colleague Branko Kolarevic the opportunity to provide his final remarks.

KOLAREVIC: I would just like to remind you where we started. Ali Malkawi and I have positioned this symposium as a fantastic territory that lies in what we describe as two poles — the pole of the conceptual, or what we call the performativity of the virtual, and the pole of the operative, or what we call the performativity of the real. I think the metaphor of the poles is an appropriate one because it relates to the sphere. And a sphere, as an endless entity, can have more than two poles, depending on how you qualify it. I think each of you were able to define your own poles, oppositional poles, in the discourse that we had over the past two days. We were not hoping to actually get some answers out of this event. As Ali Malkawi stated, our intent was to actually generate more questions than the answers. In that respect, I think, we have succeeded.

EPILOGUE

ALI M. MALKAWI

EPILOGUE

ALI M. MALKAWI

This book presents varying views on the concept of performance as it relates to buildings. Engineering and architectural perspectives, theoretical interpretations, as well as ideas from research and development, were provided by scholars, researchers and practitioners. These discussions suggest that "performance" is understood differently by the different participants in the design of buildings.

A driving force behind this assemblage was the desire to investigate the impact of the recent surge in computational instrument development and use in architectural practice and "high performance" buildings. Although the performance concept is not new, and is inherently a basic component in building design, its application to the building envelope and structure that is represented in "high performance" buildings is relatively recent.

The book illustrates that buildings embodying the notion of "integration" or "high performance" concepts are developed through highly interactive relations between architects and engineers. This integration is influenced primarily by the "process" of collaboration. This process has been imbued by advancements in computational technology that narrowed the gap between the design actors, including architects and engineers. It made it easier for engineers to be engaged in all aspects of design activities, and be able to investigate and clarify questions by conducting fast, robust analysis. This allowed new forms of architecture to be possible.

For buildings that utilize "integration" or "performance" as their main theme, the process of having the designers orchestrate the design integration principles, and having the instruments bridge the gap between the participants, has proven to be essential.

Currently, computational instruments are used for analysis to support design decisions. They are used to check solutions and to convince clients. The architects typically integrate performance principles early in the design process. These principles affect both the form and the structure of the building, and they get to be refined and optimized by the engineer's active participation within the process using such instruments. Issues that used to be difficult to predict are now possible within a relatively short period of time.

Although the profession has access to sophisticated analysis tools, work is still needed in many areas to facilitate better integration between architects and engineers, as well as among the engineers themselves. Building professionals have not investigated the full potential of computation to aid in the opening of new horizons for building performance. Integration of the advancement in computational theory and building simulation is still in its infancy. The problem is complicated by the fact that the development of these issues is driven by "immediate" market needs rather than long-term goals. New commercial instruments are being developed in an attempt to respond to users' needs. These instruments still lag in their progress due to a "disconnect" between funded efforts and commercial tool developments.

The book illustrates that, although the instruments have been maturing, only elite practices are utilizing them. These practices have shown that these instruments can be a major influence on the way buildings are conceived, designed and constructed. Although relevant chapters of the book illustrate how technological developments in computational analysis make it possible to bridge the gap between the engineering and architectural realm, much more can be accomplished with a focus on the development of tools that will provide design synthesis rather than only analysis. The multi-criteria and complex nature of building performance predictions illustrates the need for more work in this area.

APPENDIX
AUTHOR BIOGRAPHIES
PHOTO CREDITS
INDEX

APPENDIX

AUTHOR BIOGRAPHIES
PHOTO CREDITS
INDEX

BIOGRAPHIES

GODFRIED AUGENBROE
Associate Professor
Georgia Institute of Technology
Atlanta, USA

Professor Augenbroe received an MSc *cum laude* in Civil Engineering from TU Delft in the Netherlands in 1975. He has spent most of his academic career in Europe, where he pioneered a commercial finite element toolbox for building simulation and managed large European Union-funded projects on engineering interoperability and semantic building product models. He has led international consortia of academic researchers and industrial developers, such as the COMBINE effort (Computer Models for the Building Industry in Europe), a project spanning 1990–95. This project has delivered the first prototypes of the next generation of integrated engineering design systems with an emphasis on building services engineering.

Professor Augenbroe has been active in the pursuit of international collaboration, exemplified in prolonged working stays abroad at CSTB, France (1988), Lawrence Berkeley Lab, USA (1989), and at UNPHU, Dominican Republic (1996). More recently he has held visiting professorships at Loughborough University, UK, and the University of Newcastle upon Tyne, UK. Since 1997 he has been heading the Building Technology track in the Doctoral Program in the College of Architecture at the Georgia Institute of Technology in the USA.

Professor Augenbroe teaches graduate courses and conducts research in the fields of building performance concepts and simulation, the control of smart systems, e-business, system monitoring and diagnostics. He has also established an active research record in web-hosted collaboration and knowledge management, dealing with the development of software tools and their business integration.

Professor Augenbroe is on the scientific board of five international journals, is American co-editor of two other scientific journals and has published over a hundred refereed papers. He has chaired three major conferences and delivered six keynote lectures at international conferences. He was the chair of the IBPSA BS2003 conference in August 2003, and he is currently the coordinator of the e-HUBs project (IST-2001-34031). The project focuses on the tactical decision-making that prepares e-engineering partnerships, which are becoming ever more vital for the effective globalization of the building industry.

http://www.coa.gatech.edu/phd

JEAN-FRANÇOIS BLASSEL
Director
RFR Consulting Engineers
Paris, France

For more than ten years, Jean-François Blassel has been one of the directors of RFR, the unique architectural engineering firm founded by Peter Rice in Paris.

Jean-François Blassel trained as an architect and an engineer, and has collaborated on many large-scale structures where architecture and technology are intertwined from the beginning of the project.

He has worked in Europe, North Africa, Asia and the United States on large-scale buildings and structures. These projects include the Kansai International Airport with Peter Rice and Renzo Piano, several high-speed train (TGV) stations and viaducts for the French railroad authority, as well as many smaller technologically-oriented projects.

Jean-François Blassel also teaches at the Ecole d'Architecture, de la Ville et des Territoires in Paris and has been a visiting faculty for the "Emerging Technologies" program at the Graduate School of Fine Arts of the University of Pennsylvania since 1999.

Jean-François Blassel received the 1994 French Government's "Album de la Jeune Architecture" Award for architects under forty, and he writes regularly for the French architectural press.

http://www.rfr.fr

WILLIAM W. BRAHAM

Associate Professor of Architecture
University of Pennsylvania
Philadelphia, USA

William W. Braham is an Associate Professor of Architecture at the University of Pennsylvania. He received a degree in Civil Engineering from Princeton University and an MArch and PhD Arch from the University of Pennsylvania, where he has taught since 1988. At Penn he teaches graduate courses on light and environmental technology, and he coordinates the second-year design studios. He practices with Studio Luxe, an architectural design and consulting practice, and he sits on the boards of Ivalo Lighting and of Praxis, Penn's design practice unit. In 2002 he published a book called *Modern Color/Modern Architecture: Amédée Ozenfant and the genealogy of color in modern architecture* (published by Ashgate). He is currently co-editing a collection called *Rethinking Technology: A Reader in Architectural Theory*, due to be published in 2005, and he is writing an architectural account of environmental conditioning since 1968.

http://design.upenn.edu/~brahamw/

JAN EDLER

Designer and Architect
realities:united
Berlin, Germany

Jan Edler studied architecture at the Technical University Aachen and at the Bartlett, University College London. He graduated in 1997 as a Diploma Architect. Since 1996 he has worked as a co-founder for the Berlin-based art group [kunst und technik]. In 2000 he founded the architectural design studio realities:united (realU) together with his brother Tim Edler.

The team deals comprehensively with the staging of cultural events and in the designing of material and information spaces. realU does research for the development of new technologies and progressive working methods, ideas, messages and communication strategies. realU employs a great variety of scientific, commercial and artistic methods and strategies — often mixing and blurring conventional discipline-related work strategies.

Since 1998 realU has been focusing on projects researching the integration of media technology in the "art space," both for self-commissioned and commercial projects. Current projects include expertise and strategic design work for the integration of media and communication technology for the Kunsthaus Graz, Austria, the new Paul Klee Museum in Bern, Switzerland, and a research project on prototypical video communication for the Bauhaus-Dessau, Germany.

Their projects were presented at numerous international exhibitions and conferences, and have been published worldwide. They were awarded several internationally recognized design distinctions.

Recent teaching positions held by the duo include the Department of Architecture at the Technical University of Berlin (2000–01), the Bauhaus Foundation, Dessau (2001–02), and the Pasadena Art Centre College for Design, Los Angeles, USA (2003).

http://www.realities-united.de

THOMAS HERZOG

Principal
Herzog + Partner
Munich, Germany

Professor Herzog has been the Dean of the Faculty of Architecture at the Technical University in Munich (TUM), Germany, since 2000, and he has been a Guest Professor at Tsinghua University in Beijing, China, since 2003. He received his Diploma from TUM in 1965 and earned his doctorate in architecture at the University of Rome "La Sapienza" in 1972 on "Pneumatic Structures."

Thomas Herzog founded his practice in 1971. Since then he has worked jointly with Verena Herzog-Loibl. His work has been focusing on the development of building systems for the use of renewable forms of energy, the development of new building products, housing, administration, industrial and exhibition buildings, etc. He was a partner with Michael Volz from 1983–89 and has been a partner with Hanns Jörg Schrade since 1994.

Professor Herzog was the Guest Professor at Ecole Polytechnique Féderal de Lausanne (EPFL) and recently the Graham Professor at the University of Pennsylvania. He is a member of many professional organizations, including the International Academy of Architecture of UNESCO, Sofia, Bulgaria, and the Academy of Sciences and Arts, St. Petersburg, Russia. He has won many honors and awards, including the Mies van der Rohe Prize in 1981, the UIA August Perret Award for Technology in Architecture in 1996, the Grand médaille d'or d'architecture, Académie d'Architecture, Paris, in 1998, and the European Prize for "Solares Bauen" in 2000. He is the author and editor of more than a dozen specialized books in seven languages, including seven monographs. He was the German General Commissioner of the International Biennale of Architecture in Venice in 2000 and has been the Expert "Deutsche Forschungsgemeinschaft" since 2000.

http://www.herzog-und-partner.de/

HARALD KLOFT

Principal
osd – office for structural design
Darmstadt, Germany

As an engineer, Harald Kloft has a passion for the close relationship between structural engineering and architecture. He promotes the idea of structural design in which structural principles are integral to the architectural design process. He defines structural design as a process "supporting the architecture" in which structures are developed and materiality is researched to convey the architecture's ideas. Such understanding of the structural design was the basis for his undertaking of the engineering leadership in the office of Bollinger + Grohmann for some challenging "non-standard" architectural projects, such as Bernhard Franken's "Bubble" and "Dynaform" pavilions for BMW, and Peter Cook and Colin Fournier's Kunsthaus Graz, Austria.

In 2002, Harald Kloft founded the office for structural design (osd) together with Sigurdur Gunnarsson, Klaus Fäth and Viktor Wilhelm. osd is an engineering practice whose work is focused on the integration of innovative aspects of structural design in architectural projects, such as experimentation with new materials and the digital control of a design process that allows simultaneous formulation and the examination of ideas. Current projects include the Klöpp Ministry Buildings in Reykjavik, Iceland (architect Bernhard Franken), a new RegioTram Railway Station within the central station in Kassel, Germany (architects Pahl/Weber-Pahl), and a new Institute Building for the German Center of Aviation and Space Travel (DLR) in Stuttgart, Germany (archtiects bk + Arno Brandlhuber).

Aside from his practice, Harald Kloft is continuously engaged in academia. He received a doctoral degree at the University of Technology in Darmstadt, Germany. Since the mid-1990s, he has lectured in many countries and has taken part in various research and teaching programs, transmitting his experiences and knowledge of practice. He is a Visiting Professor at the Städelschule in Frankfurt, in the architectural class of Ben van Berkel and Johan Bettum. Since 2002 he has held a tenured professorship at the University of Kaiserslautern, Germany, where he chairs the Department of Structural Design at the Faculty of Architecture, and where he is building up a research program in the field of "materials and form in architecture."

http://www.o-s-d.com

BRANKO KOLAREVIC
Associate Professor of Architecture
University of Pennsylvania
Philadelphia, USA

Branko Kolarevic joined the University of Pennsylvania in 1999, where he teaches design and digital media courses, such as "Digital Morphogenesis" and "Digital Fabrication." Prior to joining Penn, he taught at universities throughout North America (Boston, Los Angeles, Miami) and Asia (Hong Kong). He has lectured worldwide on digital media in design, most recently on the "virtual design studio," "relations-based design" and "digital architectures." In 2000, he founded the Digital Design Research Lab (DDRL) at Penn.

He has published extensively in the proceedings of ACADIA, CAADRIA and SIGRADI, and has edited and authored several books, including *Architecture in the Digital Age: Design and Manufacturing* published by Spon Press, London. He is the Review Editor in Architecture for the *Automation in Construction* journal, published by Elsevier Science Publishers, Amsterdam. He is the past President of the Association for Computer Aided Design in Architecture (ACADIA). In 1998 he chaired ACADIA's organizing committee of the first Internet-based design competition for the "Library for the Information Age."

He received Doctor of Design (1993) and Master in Design Studies (1989) degrees at Harvard University, Graduate School of Design. He also holds the Diploma Engineer of Architecture degree (1986) from the University of Belgrade, Faculty of Architecture.

http://www.gsfa.upenn.edu/ddrl/

DAVID LEATHERBARROW
Professor of Architecture
University of Pennsylvania
Philadelphia, USA

David Leatherbarrow is Professor of Architecture and Chairman of the PhD program at the University of Pennsylvania, where he has taught architectural design and theory since 1984, and where he was Departmental Chair between 1992 and 1998. Before going to Penn he taught in England at the University of Cambridge and at the Polytechnic of Central London. He has also visited and taught at many universities in the USA and abroad. David Leatherbarrow studied architecture at the University of Kentucky, where he earned his Bachelor of Architecture degree, and he completed research for his PhD in Art at the University of Essex, UK.

In his scholarly work he has published a number of books, the most recent being *Topographical Stories: Studies in Landscape and Architecture* (2004). *Surface Architecture* (MIT Press, 2002), written in collaboration with Mohsen Mostafavi, won the CICA International Book Award: The Bruno Zevi Prize. Earlier books include *Uncommon Ground: Architecture, Technology and Topography* (MIT Press, 2000), *The Roots of Architectural Invention: Site, Enclosure and Materials* (Cambridge University Press, 1993), and, also with Mostafavi, *On Weathering: the Life of Buildings in Time* (MIT Press, 1993), which won the 1995 International Book Award in architectural theory from the American Institute of Architects. These books have been frequently reviewed in scholarly journals. Additionally, he has published over fifty scholarly articles in architectural journals, including *AA Files, Architectural Design, Center, Daidalos, Journal of Garden History, Journal of the Society of Architectural Historians, Rassegna, Via,* and others.

His awards include a Visiting Scholars Fellowship at the Centre Canadien d'Architecture, the Cass Gilbert Distinguished Professorship at the University of Minnesota, and a Fulbright Hays Scholarship for study in Great Britain. In the past, his research has focused on topics in the history and theory of architecture, gardens and the city; more recently his work has concentrated on the impact of contemporary technology on architecture.

http://www.design.upenn.edu

ALI MALKAWI

Associate Professor of Architecture
University of Pennsylvania
Philadelphia, USA

Ali Malkawi joined the University of Pennsylvania in 2001 as an Associate Professor. He teaches architectural technology and computation in the Master of Architecture program. He also conducts research in the areas of computational simulation and building performance evaluation. In addition, he supervises research and teaches in the PhD program. He was the founder and former coordinator of the Augmented Reality Group at the University of Michigan, and he currently heads the Building Simulation Group at Penn. He taught and conducted research at the Georgia Institute of Technology, the University of Michigan and Harvard. He has lectured at numerous universities and conferences, and has been published in many proceedings and journals.

Dr. Malkawi received his PhD in Architecture and Artificial Intelligence in 1994 from the Georgia Institute of Technology. He also holds a MArch degree from the University of Colorado and a BS in Architecture Engineering from the Jordan University of Science and Technology.

http://pobox.upenn.edu/~malkawi

PETER McCLEARY

Professor of Architecture
University of Pennsylvania
Philadelphia, USA

Professor McCleary studied and graduated in applied mathematics and natural philosophy at Glasgow University, civil engineering at the University of Strathclyde, structural engineering at Imperial College, London, and architecture at the University of Pennsylvania.

He has practiced engineering and architecture in London with Ove Arup, Arup Associates, Felix Samuely (with Frank Newby) and Freeman-Fox. He has had an architectural and engineering consulting practice since 1965. Among those architects and engineers that he has consulted with are: Lewis Davis, Bernard Huet, Louis Kahn, Jean-Marc Lamuniere, Ian McHarg, and Robert Le Ricolais.

Professor McCleary was invited to the University of Pennsylvania in 1965, where he served as Chairman of the Masters and PhD programs in Architecture, and he was the founder and first Chairman of the Historic Preservation program. He teaches structures, aesthetics of bridges, philosophy of technology, the relationship between structure and space, and design studio.

He is an invited Professor of Technology and Design at many universities in the US, Canada, Europe, Australia and the Middle East. He has received research grants from NEA, the National Trust for Historic Preservation, the French Ministry of Culture, and Fellowships from NEA and the Fulbright and Graham Foundations. He was the founder of the ACSA Annual Technology Conference, and was awarded their Distinguished Professor Medal.

Professor McCleary continues to write many research and interpretive articles on the relationship between the technologies of architecture and engineering, which are published in several languages in numerous journals and books.

http://www.design.upenn.edu

ALI RAHIM
Assistant Professor of Architecture
University of Pennsylvania
Philadelphia, USA

Ali Rahim is an Assistant Professor of Architecture in Media and Design at the University of Pennsylvania and he is Director of the Contemporary Architecture Practice in New York City. He has established an award-winning profile in futuristic work using digital design and production techniques.

He was the recipient of the Honor Award for Excellence in Design from Columbia University where he received his Master of Architecture. Current design research includes a range of residential, commercial and product design projects. His architectural designs have received awards and have been exhibited widely. They include a winning competition entry for Ephemeral Structures for the 2004 Olympic Games, a shopping mall, steel museum and a one-acre naval memorial. Past exhibition venues include the Artists Space Gallery in New York City, and the Royal Institute of British Architects (RIBA) in London. His most recent exhibition was in Delft, the Netherlands. He has also lectured extensively about his work in Asia, Europe and North America.

His guest edited books and journals include *Contemporary Techniques in Architecture* (Academy Editions/ Wiley and Sons, February 2002) and *Contemporary Processes in Architecture* (Academy Editions/Wiley and Sons, August 2000). His projects and articles have been published by Actar Press (Barcelona), Columbia University Press (New York City) and MIT Press (Cambridge, Mass.), among others. Currently he is working on a monograph of his work, *Manual for Digital Media and Architecture Design* (Academy Editions, London, forthcoming 2004), and a special issue for *Architectural Design* entitled "Contemporary Cultural Production in Architecture" (Academy Editions/Wiley and Sons, forthcoming 2005).

http://www.c-a-p.net

MAHADEV RAMAN
Principal
Arup
New York, USA

Mahadev Raman leads Arup's Building Engineering Group in New York and is a member of the firm's Americas Board. He has been with Arup since 1978, providing engineering design leadership for multidisciplinary teams on a wide variety of projects worldwide. Prior to joining the New York office, he worked in Arup's London and Cambridge offices.

He brings particular expertise in the design of sustainable, high performance and energy-efficient buildings. He has pioneered the use of sophisticated analytical techniques to improve the performance of low-energy designs.

Mahadev Raman lectures regularly at Princeton and Columbia Universities on sustainable design and environmental control. He has a Bachelor of Engineering Science from the University of Durham, UK (1978) and a Master of Science in Applied Energy from the Cranfield Institute of Technology, UK.

http://www.arup.com

CRAIG SCHWITTER
Partner
Buro Happold
New York, USA

Craig Schwitter is one of the youngest partners in Buro Happold. He established the New York office in 1998 after gaining overseas experience working on the Millennium Dome in London and various other projects in the UK. He has led both the structures and MEP groups on key projects, such as Santiago City of Culture, Rensselaer Polytechnic Institute and the new green Genzyme Headquarters.

His experience in the Buro Happold UK offices motivated him to seek to bring the best aspects of the European collaborative team design approach to the United States. In setting up the New York office, his vision is to utilize the fullest depth of Buro Happold's international practice knowledge while recognizing the importance of mobilizing locally-based experience to ensure a project's success. Under Craig Schwitter's guidance, and along with his belief that excellent architecture and excellent engineering are inseparable, the New York office quickly earned a reputation for outstanding, creative engineering. As a result, the office expanded rapidly to the current staffing level of around thirty people.

In 2002, he was appointed as a Trustee for the Van Alen Institute, which is committed to improving the design of the public realm.

http://www.burohappold.com

LARS SPUYBROEK
Principal
NOX Architects
Rotterdam, Netherlands

Since the early 1990s Lars Spuybroek has been involved in researching the relationship between architecture and media, often more specifically between architecture and computing. He was the editor and publisher of one of the first magazines in a book format (*NOX*, and later *Forum*), he made video (*Soft City*) and interactive electronic artworks (Soft Site, edit Spline, deep Surface), with the last five years spent focusing more on architecture (HtwoOexpo, V2_lab, wetGRID, D-Tower, Son-O-house, Maison Folie).

His work has won several prizes and has been exhibited all over the world, including presentations at the Venice Biennale in 2000 and 2002. He is currently working on an interactive tower for the Dutch city of Doetinchem (D-Tower), "a house where sounds live" (Son-O-house), an interactive office building in Stratford-upon-Avon, England (SoftOffice), a multiple complex of cultural buildings in Lille, France, and a Home for Alice, in Italy.

Lars Spuybroek has lectured all over the world and has taught at several universities in the Netherlands. He is a regular Visiting Professor at Columbia University. Since 2002, he has held a tenured professorship at the University of Kassel in Germany where he chairs the CAD/digital design techniques department. His book *Machining Architecture* will be published by Thames & Hudson in 2004.

http://www.noxarch.com

ANDREW WHALLEY
Director
Grimshaw
London and New York

Andrew Whalley has worked at Grimshaw since 1986. He works across a broad range of sectors and has undertaken projects in a number of countries which include: Donald Danforth Plant Science Center in St. Louis, USA; Southern Cross Station, Melbourne, Australia; Fundacion Caixa Galicia, Coruna, Spain; Paddington Station, London; the Eden Project, Cornwall, UK; the Royal College of Art in London; and the new spa in the city of Bath, UK.

In addition to working on projects, he holds the responsibility in the office for developing new business and for overseeing publicity and publications. He also acts as a visiting lecturer at the Bartlett, University College London, and as a visiting professor at Washington University in St. Louis. He has recently spoken at events and symposia in Italy, the USA, Norway, Germany, Holland, and at various locations in the UK.

Andrew Whalley studied at the Mackintosh School of Architecture in Glasgow and at the Architectural Association in London. He is registered as an architect in the United Kingdom and in the United States.

Andrew also works in partnership with his wife, Fiona Galbraith, and together they have designed a house in Dollar, which won an RIBA Award for Scotland, and more recently they have completed a house and studio for themselves in London. He is based principally in Grimshaw's New York office and is overseeing the recent competition-wining project, the Fulton Street transit center. Andrew Whalley has written a book with Hugh Pearman entitled *The Architecture of Eden*, which was published by Transworld in October 2003.

http://www.grimshaw-architects.com

PHOTO CREDITS

COVER Peter Macapia

1

1.1–14 Anna Catherine Vortmann Hakes, except:
1.6 David Leatherbarrow
1.11 David Leatherbarrow

2

2.1 Michael Dyer Assoc.
2.2 Michael Dyer Assoc.
2.3 Grimshaw
2.4 Jo Reid and John Peck
2.5a Andrew Whalley/Grimshaw
2.5b Peter Stroebel
2.6 Peter Cook/VIEW
2.7 Peter Stroebel
2.8 Jo Reid and John Peck
2.9 Werner Huthmacher
2.10 Jens Willebrand
2.11a Andrew Southall
2.11b Edmund Sumner
2.12 Peter Cook/VIEW
2.13 Grimshaw
2.14 Grimshaw
2.15 Michael Dyer Assoc.
2.16 Timothy Hursley
2.17 Grimshaw
2.18 Grimshaw
2.19 Jens Willebrand
2.20 Nathan Willock
2.21 Grimshaw
2.22 Edmund Sumner
2.23 Grimshaw
2.24 Grimshaw
2.25 Edmund Sumner
2.26 Grimshaw
2.27 Grimshaw
2.28a Grimshaw
2.28b Grimshaw
2.29a Grimshaw
2.29b Grimshaw/Univ. of Bath/UCL/Bentley
2.29c Grimshaw/Univ. of Bath/UCL/Bentley
2.30 Grimshaw
2.31a Grimshaw
2.31b Grimshaw

3

3.1–17 Arup, except:
3.2 BNIM Architects
3.5 Andy Ryan
3.11 Frederick Charles
3.12 Frederick Charles
3.13 Peter Mauss/ESTO
3.14 Peter Mauss/ESTO
3.17 Yutaka Nagata © United Nations UN/DPI

4

4.1 Architectural Record/McGraw-Hill
4.2 Productivity Press
4.3 Productivity Press
4.4 Greg Lynn/Princeton Architectural Press
4.5 Regents of the Univ. of California
4.6 Reinhold Publishing Corp.
4.7 Sweets Construction/McGraw-Hill
4.8 Sweets Construction/McGraw-Hill
4.9 Ottomar Gottschalk/Verlag Schnelle
4.10 Ottomar Gottschalk/Verlag Schnelle
4.11 Ottomar Gottschalk/Verlag Schnelle
4.12 Architectural Review
4.13 Van Nostrand Reinhold Company
4.14 Architectural Record/McGraw-Hill, Progressive Architecture
4.15 Ante Prima, Paris/Birkhauser, Basel

5

5.1–34 Herzog + Partner, except:
5.3 Peter Bonfig
5.4 Klaus Kinold
5.7 Dieter Leistner
5.8 Dieter Leistner
5.9 Verena Herzog-Loibl
5.10 Dieter Leistner
5.11 Dieter Leistner
5.12 Peter Bartenbach
5.13 Verena Herzog-Loibl
5.15 Dieter Leistner
5.16 Verena Herzog-Loibl
5.18 Verena Herzog-Loibl
5.19 Verena Herzog-Loibl
5.20 Reinhard Demuss
5.22 Dieter Leistner
5.23 Dieter Leistner
5.25 Dieter Leistner
5.27 Dieter Leistner
5.28 Dieter Leistner
5.31 Dieter Leistner
5.34 Moritz Korn

6

6.1–7 Ali Malkawi, except:
6.1 Glenn Forney/NIST
6.5 IAI

7

7.1–7 Godfried Augenbroe

8

8.1–25 Buro Happold, except:
8.10 Buro Happold/Adam Wilson
8.17 Foster and Partners/Nigel Forster
8.18 Foster and Partners/Nigel Forster
8.20 Mandy Reynolds/Fotoforum
8.21 Mandy Reynolds/Fotoforum
8.22 Mandy Reynolds/Fotoforum

9

9.1–12 RFR

10

10.1 Harald Kloft
10.2 Harald Kloft
10.3 Irving Penn © Austrian Frederick and Lillian Kiesler Private Foundation
10.4 Fritz Busam
10.5 ABB/Bernhard Franken
10.6 ABB/Bernhard Franken
10.7 ABB/Bernhard Franken
10.8 Bollinger + Grohmann
10.9 Bollinger + Grohmann
10.10 Harry Schiffer
10.11 spacelab.uk
10.12 Bollinger + Grohmann
10.13 Bollinger + Grohmann
10.14 Bollinger + Grohmann
10.15 Bollinger + Grohmann
10.16 Bollinger + Grohmann
10.17 Bollinger + Grohmann
10.18 Bollinger + Grohmann
10.19 Franken Architekten

10.20a osd – office for structural design
10.20b osd – office for structural design
10.21 ABB/Bernhard Franken
10.22 Harald Kloft
10.23 Bollinger + Grohmann
10.24 Harald Kloft
10.25a Harald Kloft
10.25b Harald Kloft
10.25c Harald Kloft
10.26 Harald Kloft
10.27 Fritz Busam
10.28 Tilman Elsner

11

11.1–16 realities:united, except:
11.1 Harry Schiffer
11.2 Colin Fournier
11.3 Colin Fournier
11.6 ArGe Kunsthaus, Graz/ Pichlerwerke GmbH, Graz
11.7 Harry Schiffer
11.11 Landesmuseum Joanneum, Graz
11.12 Harry Schiffer
11.16 Landesmuseum Joanneum, Graz

12

12.1–9 NOX/Lars Spuybroek

13

13.1–9 Ali Rahim/Contemporary Architecture Practice

14

14.1 Bernhard Franken/ABB Architekten
14.2 Bernhard Franken/ABB Architekten
14.3 Arup
14.4 Future Systems
14.5 Bollinger + Grohmann
14.6 Arup
14.7 Gehry Partners
14.8 Kristina Shea
14.9 Kristina Shea
14.10 Matthew Herman/Univ. of Pennsylvania

15

15.1 NOX/Lars Spuybroek
15.2 NOX/Lars Spuybroek
15.3 NOX/Lars Spuybroek
15.4 NOX/Lars Spuybroek
15.5 Colin Fournier
15.6 Landesmuseum Joanneum, Graz
15.7 Graeme Peacock
15.8 Gifford and Partners
15.9 Colleen and Ted Garringer
15.10 Alan Karchmer/ESTO
15.11 Mark Burry and Mark Goulthorpe/dECOi
15.12 Foster and Partners
15.13 Renzo Piano Building Workshop
15.14 Foster and Partners
15.15 Arup

16

16.1–3 Arup

INDEX